Iwanitz · Lange

OPC

OPC

Fundamentals, Implementation, and Application

3rd rev. Ed.

Frank Iwanitz • Jürgen Lange

UNIVERSITY SCIENCE PRESS

(An Imprint of Laxmi Publications Pvt. Ltd.)

BANGALORE • CHENNAI • COCHIN • GUWAHATI • HYDERABAD
JALANDHAR • KOLKATA • LUCKNOW • MUMBAI • RANCHI
NEW DELHI

Published by :
UNIVERSITY SCIENCE PRESS
(*An Imprint of Laxmi Publications Pvt. Ltd.*)
113, Golden House, Daryaganj,
New Delhi-110002

Phone : 011-43 53 25 00
Fax : 011-43 53 25 28

www.laxmipublications.com
info@laxmipublications.com

Originally published 2005 in German language under the title Frank Iwanitz, Jürgen Lange: OPC. Grundlagen, Implementierung und Anwendung.

Indian Edition 2010

This special low-priced edition is only for sale in India, Nepal, Sri Lanka, Bangladesh and Pakistan.

Price : **Rs. 295.00** *Only.* *Edition* : 2010

OFFICES

© **Bangalore**	080-26 61 15 61	© **Chennai**	044-24 34 47 26
© **Cochin**	0484-237 70 04, 405 13 03	© **Guwahati**	0361-254 36 69, 251 38 81
© **Hyderabad**	040-24 65 23 33	© **Jalandhar**	0181-222 12 72
© **Kolkata**	033-22 27 43 84	© **Lucknow**	0522-220 95 78
© **Mumbai**	022-24 91 54 15, 24 92 78 69	© **Ranchi**	0651-221 47 64

UOP-9468-295-OPC FUND IMPL & APP-IWA **C— 951 /010/01**

Typeset at : Kalyani Computer Services, New Delhi. *Printed at* : Printed and bound in India by Print O Pack, Delhi

Acknowledgements

Since December 1997, Softing AG has been running seminars on OPC. In this time, the amount of seminar documents has increased, and knowledge on the subject has become more and more complex. In 1999 the idea arose to make this knowledge available to a larger public in the form of a book. It was a long journey from the first steps of realizing this idea to the completion of this book. Many thanks to all those who have assisted us.

First of all there are the countless participants in our seminars, who – because of their questions and comments – have drawn our attention to many aspects of OPC which we had not realized, and who gave us incentives to present the subject of OPC in even more concrete terms.

The subject of OPC is treated by a large number of staff with Softing AG. Without their support and their willingness to pass on their knowledge, it would not have been possible to write this book. On many occasions they put themselves at our disposal. Therefore, we would like to thank our former and current colleagues David Jenny, Peter Datsichin, Michael Meisinger and Thomas Rummel.

Many thanks also to all our other colleagues within Softing who have contributed to improving the contents of this book by giving us advice and hints.

Many thanks also to the authors of the articles for chapter 4, who – by means of their application descriptions for OPC and their user reports – have contributed an additional illustration of this technology.

Not least, we would like to thank the publishing house Hüthig, in particular Mr. Schultz, who assisted us in realizing this book.

Finally we want to thank Tony Prylowski, Ian Pepper and David Handy of Automsoft International who helped us a lot to improve the English translation of this book.

A lot of positive feedback from readers, and the fact that the first and second editions sold more than five thousand copies have motivated us to work on this third edition. Thanks to all the readers who helped us to improve the book by their comments.

Haar, October 2005

Frank Iwanitz (frank.iwanitz@softing.com)
Jürgen Lange (juergen.lange@softing.com)

During the work on this book I received much support and encouragement from my family. Without their understanding and patience it would not have been possible to finish the work. Therefore my special thanks go to my wife Tatjana and my daughters Marianne and Kristina.

Munich, October 2005
Frank Iwanitz

The past few months were clearly marked by work on this book. Because of my absence at many evenings and weekends, I demanded a lot of my family.

Thanks to my wife Steffi and my children Stefan, Kathrin, Daniel and Christine, who exercised patience and understanding for my absence during my work on this book.

Munich, October 2005
Jürgen Lange

Preface

I am pleased to introduce the third edition of the OPC book written by my friends and colleagues Juergen & Frank. This is the real "OPC Book" that has become the fundamental reference manual and overview for the OPC foundation technology providing a real foundation of knowledge for anyone that wants to become literate on the OPC foundation technology. Whether you are an end user of products from vendors that use the OPC foundation technology or you are a vendor that either currently builds products with the OPC foundation technology or potentially will in the future, then this is the book for you to truly get jump started on understanding the OPC technology and using the OPC technology in a highly efficient manner.

I have had the pleasure of knowing the authors of this book on both a personal and professional basis.

The authors have been active participants in the development of the OPC technology since it's initial conception. The authors have written this book to truly provide the foundation of knowledge on the technology and communicate the total picture of OPC from the beginning to where OPC is today.

The book provides an overview of the history of the OPC Foundation. It describes the advantages of the OPC Technology and gives detailed descriptions of the OPC specifications and the current status of the work of the technical subcommittees. It gives developers practical guidelines on the implementation and use of the OPC specifications.

The book communicates more than just the what about OPC, it tells the reader how to design and implement products that leverages the hidden tricks of OPC in addition to the standard features. It explains the real guidelines/rules to build OPC products that maximizes the use of the technology, and provides a reference point for developers to provide leading edge products.

The authors are both developers and evangelists of the technology. The authors have in depth knowledge of how others have built products using the OPC technology.

The book describes the process for the design and implementation of OPC components and illustrates how to use OPC with application examples. It explains OPC compliance testing, security aspects, and use of OPC in the Internet. It presents an overview of OPC products in the market place.

The third edition addresses interoperability from both a compliance and interoperability perspective as well as providing the reference and understanding about the OPC technology being the "HOW" and moving "WHAT", with emphasis on moving information from the factory floor all the way to the enterprise and how the OPC foundation is collaborating with other standards organizations for reliability and interoperability beyond the manufacturing domain.

The OPC foundation has undertaken a significant architecture endeavor called the OPC unified architecture which provides a common base of services focusing on the use of Web services and XML and embracing the standardization of all the OPC foundation existing technology and specifications. The unified architecture provides services targeted at reliability and security for moving information from the factory floor to the enterprise and leverages the standards from other organizations addressing Web services interoperability. A common base of services have been defined that will allow the functionality of OPC data access, OPC alarm and events, OPC historical data access, OPC batch, OPC data eXchange and OPC commands to truly be integrated together in a highly robust and secure fashion. Vendors will now be able to support a single set of services that is the foundation for interoperability for the other OPC foundation domain specifications. Interoperability via a Web services architecture allows for true platform portability as well as being scalable to address the different deployment technology that is needed to build products in all facets of automation. Performance, security and reliability are the attributes of the OPC unified architecture. Existing OPC products will be able to take advantage of the unified architecture by essentially plugging in and leveraging the OPC unified architecture wrappers. This will allow end-users currently using products that are based on OPC foundation DCOM based technology to take advantage of the services and reliability provided by the OPC unified architecture without requiring a product upgrade from the vendor's for the existing products that are currently in use. Native adoption of the OPC unified architecture will be retrofitted into the existing products supplied by the vendor's further enhancing the reliability and interoperability that the end-users expect for true automation interoperability.

Thanks again to the authors for putting together such an excellent book that communicates the whole picture of OPC, and provides a real world guidelines for developers and end users to understand and develop OPC components that takes advantage of the OPC technology.

Thomas J. Burke
President & Executive Director OPC Foundation

Foreword

The pace of change in industrial control and automation technology is accelerating. Demands on machines and systems concerning flexible retrofit, production speed, and fail safety are increasing, as are cost pressures. Software is becoming more and more the essential factor with products, systems, and complete plants.

At the same time, changes in the field of automation brought by the use of the PC as an automation component, by the Internet, and by the tendency to more open standards, can be clearly seen – to the benefit of the user and the manufacturer.

The PC is more and more used for visualization, data acquisition, process control, and the solution of further tasks in automation. It complements or replaces the traditional PLC and the operator terminal. The reasons for this are the continual decrease in price of the mass-produced PC, the permanent multiplication of the computing capacity of the CPU, the availability of even more efficient and comfortable software components, and the ease of integration with Office products.

Efficiency and cost savings are achieved through the reuse of software components and the flexible compilation of such components into distributed automation solutions.

Horizontal integration of the automation solutions through communication between the distributed components also plays an important role. Immense additional savings are achieved by means of vertical integration by optimizing the process of product planning, development, manufacturing and sales. This optimization is realized through a consistent data flow, permanent data consistency, and the availability of data on the field level, control level, and office level.

The use of standardized interfaces by several manufacturers is a prerequisite for the flexible compilation and integration of software components.

OPC is now generally accepted as the most popular industrial standard among users and also among developers. Most of the HMI (Human Machine Interface), SCADA (Supervisory Control and Data Acquisition), and DCS (Distributed Control System) manufacturers in the field of PC-based automation technology, as well as the manufacturers of Soft PLCs, are offering OPC client and/or OPC server interfaces with their products. The same is true for suppliers of devices and interface cards. In the last years OPC servers widely replaced DDE (Dynamic Data Exchange) servers and product-specific drivers in this field.

Today, OPC is the standard interface for access to Windows-based applications in automation technology. Over 7,000 OPC products are in use worldwide in a vast variety of industries. The number of installations has reached many hundred thousands and covers a wide range of application areas. The OPC interface is today not only used for simply transmitting process data or individual parameters: Entire ERP documents, parameter sets, control sequences, video signals or drive programs are transferred via OPC. Within only a few years, OPC has conquered the industry. Whether in the production industry, in process technology or in building automation, OPC is the standard! This fact is substantiated by a study with which the OPC Foundation commissioned the ARC market research company in 2003. The result of the ARC study states that 84% of all participants in the study are very familiar with OPC technology. The study adds that in 78% of the production systems and MES applications, 75% of the HMI/SCADA systems, 68% of the process supervisory control systems and PLCs, and 53% of the ERP systems, OPC is the preferred method for data exchange and connection to the process. The market research company expects this trend to continue at an even greater pace over the next five years, with OPC extending to subsystems, small automation components, and even down to the sensor level. OPC, as implemented in the products available today, is solely based on the Distributed Component Object Model (DCOM), Microsoft's technology for the implementation of distributed systems. Since Microsoft's announcement that development of DCOM will be discontinued and superseded by .NET, completely new concepts for OPC have been developed, which protect the investments in existing OPC developments, but will open up whole new application possibilities for OPC. In future, more and more data will be exchanged via Web Services, and the eXtended Markup Language (XML) will be increasingly utilized as a description language. OPC specifications define an interface between clients and servers as well as servers and servers for different fields of application – access to real time data, monitoring of events, access to historical data, and others. Just as any modern PC can send a print task to any printer, thanks to the integration of printer drivers, software applications can have access to devices of different manufacturers without having to deal with the distinct device specifications. OPC clients and servers can be combined and linked like building blocks by the use of OPC technology.

At present, OPC clients and servers are mainly available on PC systems with Windows 9X/ ME/NT/2000/XP and x86 processors. As the current Windows CE versions support DCOM, various solutions (operator panels, handhelds, etc.) provide an OPC client or server interface. Due to the portability of Microsoft's .NET framework, and particularly the availability of Web services for multiple operating systems, the use of OPC components in different environments and in embedded systems will become more important in future; first products are already available.

Why is OPC so successful?

When creating OPC specifications, the main focus has been and still is the practically feasible and not the theoretically desirable. The main center of attention is the 60% of functionality required by 80% of users. The approach of the OPC Foundation has always been to avoid unnecessarily detailed discussions and political disputes and to create practical facts within a very short time. The Foundation therefore succeeds again and again to build additional requirements into the specifications, to rapidly develop new OPC specifications, and to successfully implement them in practice.

Although CORBA (Common Object Request Broker Architecture) and EJB (Enterprise Java Beans) are comparable technologies to the DCOM, no specifications comparable to OPC and based on these technologies exist. OPC has succeeded in defining a uniform standard worldwide which was adopted by manufacturers, system integrators, and users in only three years.

The book on hand is addressed to all those interested in OPC and who want to know more about it – users of OPC products, decision-makers, product managers, project managers, and software developers. This revised, third edition provides for "beginners" and experienced OPC users alike the opportunity to become familiar with the current status of the OPC specifications as well as important news, for example, OPC Unified Architecture (UA) or various collaborations of the OPC Foundation with different other standardizations organizations.

The reader should have a general understanding of the requirements of modern automation technology. Basic knowledge of programming with object-oriented languages would be an advantage. Technical terms from the fields of automation technology and software development, which may be less common, are explained the first time they appear in the book.

In chapter 1 of this book, the history of OPC is described. Technological development – from specific drivers limited to one product, via the first standardization efforts in the form of DDE, up to the OPC technology – is described. The advantages and disadvantages to the development and use of product specific drivers, DDE servers, and OPC servers are examined. A general overview of the basic DCOM and .NET technology and the principal functions of OPC clients and OPC servers are also included in this chapter. The chapter finishes with an insight into the structure and operation of the OPC Foundation.

Chapter 2 explains the existing OPC specifications and those being developed, such as Common Definitions and Interfaces, Data Access, Alarms and Events, Historical Data Access, Batch, Security, Compliance Test, Complex Data, OPC XML-DA, OPC DX, Commands, and OPC Unified Architecture.

For every specification, the field of application and the model this specification is based on are presented. Afterwards, the contents and use of the specification are described using specific scenarios. This chapter finishes with an application example.

Chapter 3 gives answers to questions arising from design and implementation of OPC components. Not only are the general aspects of the use of DCOM resp. .NET explained, but also how to create OPC components with C++, Visual Basic resp. Visual Basic.NET, and using OPC toolkits. Furthermore, this chapter looks at security aspects as well as the use of OPC with different embedded operating systems.

The end users of the OPC technology are the main focus of chapter 4. It describes the use of OPC in automation products of different manufacturers. This description is completed by company reports on experience when using OPC technology. This chapter also contains surveys on the performance of OPC.

Chapter 5 contains a summary and an outlook on the future of OPC.

Content

1 Introduction

1.1 History of OPC

Since reusable software components made their entry in automation technology and replaced monolithic, customized software applications, the question of standardized interfaces between components has increased in significance. If such interfaces are missing, every integration is connected with cost-intensive and time-consuming programming supporting the respective interface. If a system consists of several software components, these adaptations have to be carried out several times.

Following the immense distribution of Windows operating systems and their coherent Win32-API in the PC area, different technologies were created to enable communication between software modules by means of standardized interfaces. A first milestone was DDE (Dynamic Data Exchange) which was complemented later on by the more efficient technology OLE (Object Linking and Embedding). With the introduction of the first HMI and SCADA programs based on PC technology between 1989 and 1991, DDE was used for the first time as an interface for software drivers to access the process periphery.

Besides the product-specific drivers for visualization, configuration, and so on, a series of product-independent DDE servers was soon created. PC plug-in cards were equipped with DDE software interfaces and thus could be operated from programs with a DDE client interface.

But DDE has had some insufficiencies, however, in particular as far as performance is concerned. This induced manufacturers of HMI programs to define some DDE extensions for increased performance. The results were different, manufacturer-specific DDE flavors, another kind of proprietary interface. Interoperability on the application level was still far away.

What are product-specific drivers?

Product-specific drivers describe software components that have been developed for a specific product. They are linked to this product so that they cannot be used with products of other manufacturers. These drivers make available data in a manufacturer-specific form.

What is DDE?

DDE stands for Dynamic Data Exchange and describes a mechanism for the exchange of data between Windows applications. If, prior to the introduction of DDE, data were only exchanged through manual "copy & paste", data exchange with DDE is effected dynamically. The information to be transferred is mapped to an item with any data type. Exchange of data is effected by transfer of messages and is carried out very slowly.

What is OLE?

OLE is the abbreviation for Object Linking and Embedding. Up to the mid-1990s, this term described the entire Microsoft technology for realizing object-oriented systems. Development of OPC began during this time.

During the development of Windows NT, the Distributed Component Object Model (DCOM) was developed as continuation of the OLE technology. Windows NT was rapidly accepted by industry. In particular, the highly expanding HMI, SCADA and DCS systems were made available for NT.

What is DCOM?

DCOM stands for Distributed Component Object Model and describes an object model for the implementation of distributed applications as per the Client-Server-Paradigm. A client can use several servers at the same time and a server can provide its functionality to several clients at the same time.

At the core of DCOM is the term *interface*. DCOM objects provide their services through interfaces. An interface describes a group of related methods (functions). Description of the interfaces defines the syntax and semantics of the services offered.

The semantics (i.e. the exact meaning of arguments, admissible return values etc.) are normally defined in the accompanying documentation (e.g. OPC specifications). Definition of syntax, however, is made in a special language, called IDL (Interface Definition Language). This language also defines how the binary presentation of such an interface should look. This means that every client can call the methods of any DCOM object in the server, irrespective of the program environments (programming language, compiler and linker version, etc.) in which both are created. The interfaces are designated by a global, clear, and 128 bit-long identifier (GUID, Global Unique Identifier). Once defined and published, DCOM interfaces are by no means to be modified. The binary compatibility of DCOM objects which is thus guaranteed is one of the basic principles of DCOM.

DCOM objects exist in DCOM servers (components). There are two principal kinds of server: DLLs, or InProc-Servers, and EXEs, or OutProc-Servers which contain the executable code of the objects. DCOM objects are identified by Class Identifiers (CLSID). The client uses the CLSID to launch a DCOM component. In doing so, it uses the functions of the DCOM library. Afterwards, the client can request interfaces from the object and can call methods at these interfaces. The location of the object is thus transparent for the client as the DCOM library takes care of this. The information necessary to carry out these tasks can be found in the registry of the operating system.

The life cycle of a DCOM object is controlled by the value of an internal reference counter. The reference counter is increased by one for every interface requested and decreased by one for every interface released by the client. The DCOM object deletes itself if the reference counter has the value 0.

DCOM also defines standard mechanisms for interactions between server and client. To do this, the client has to implement a DCOM object. The server then uses methods at interfaces of this object – Client and Server exchange their roles. Within the scope of OPC and also in this book, this mechanism is called *callback*.

A more detailed description of DCOM can be found in section 3.1.

The development of DCOM components requires challenging know-how. DCOM and firewall security mechanisms exclude each other, so DCOM cannot be used straightforwardly over the Internet. For these and other reasons, Microsoft developed a new, Internet-aware concept called ".NET" (pronounced "Dot Net").

What is .NET?

With .NET, Microsoft provides a new paradigm for software development on Windows platforms.

Based on modern concepts, the .NET framework provides an effective platform for the development of Internet applications, Web services, and desktop applications. It offers significant advantages for the developer, user, and systems integrator.

The core of .NET is its Common Language Infrastructure (CLI). The CLI is a standardized runtime environment for any language. This specification includes a common standard for (object oriented) programming languages – the Common Language Specification (CLS), and its Common Type System (CTS).

The principle is similar to Java: a CLI aware compiler for a language does not build processor-specific code, but a kind of intermediate code called the MSIL (Microsoft Intermediate Language). This code is loaded from the runtime environment (Common Language Runtime, or CLR) and transferred to a Just In Time Compiler (JIT) which creates the processor specific code at runtime.

Before the processor-specific code is created, the MSIL code is checked by the CLR: during the loading of a class, the code is verified. Also, the CLR controls access to external resources like files and network links, and releases resources no longer in use (Garbage Collection).

This concept allows the collaboration of components regardless of the language they have been developed with: for example, a class can be defined in C# and be derived in VB.

The .NET framework provides a very powerful class library and strongly supports software development. Further information is given in section 3.3.

What are Web services?

The worldwide web is increasingly used for communication between applications.

The available program interfaces of the applications are called Web services.

Due to the great variety of different hardware platforms and operating systems on which these applications run, and due to the different programming languages in which these applications are developed, no general assumptions can be made about the activation and deactivation of the applications and the implemented object models.

To allow developing the applications in such a way that they are interoperable and independent of each other, two aspects have been specified, however:

- Applications communicate via the XML protocol (SOAP) and
- Programming interfaces are described with the Web Service Description Language (WSDL).

OPC uses the aspects of platform independence and Internet capability to meet the requirements for an expansion of the implementation and application basis for OPC products.

With the increased distribution of their products and the growing number of communication protocols and bus systems, software manufacturers faced more and more pressure to develop and maintain hundreds of drivers. A large part of the resources of these enterprises had to be set aside for development and maintenance of communication drivers.

In 1995, the companies Fisher-Rosemount, Intellution, Intuitive Technology, Opto22, Rockwell, and Siemens AG decided to work out a solution for this growing problem, and they formed the OPC Task Force. Members of Microsoft staff were also involved and supplied technical assistance.

The OPC Task Force assigned itself the task to work out a standard for accessing real-time data under Windows operating systems, based on Microsoft's (OLE/)DCOM technology *OLE for Process Control*, or OPC[1]. The members of the OPC Task Force worked intensively, so that already in August 1996 the *OPC Specification Version 1.0* was available. In September 1996, during the ISA Show in Chicago, the OPC Foundation was established; it has been coordinating all specification and marketing work since then.

An important task of the OPC Foundation is to respond to the requirements of the industry and to consider adding them as functional extensions of existing or newly-created OPC specifications. The strategy is to extend existing specifications, to define basic additions in new specifications, and to carry out modifications with the aim of maximum possible compatibility with existing versions. In September 1997, a first update of the *OPC Specification* was published in the form of version 1.0A. This specification was no longer named "*OPC Specification*" but *Data Access Specification*. It defined the fundamental mechanisms and functionality of reading and writing

1 The original name "OLE for Process Control" has been replaced by "Openness, Productivity & Collaboration" to reflect the fact that OPC is no longer based on Microsoft's OLE technology but on DCOM and Web services.

process data. This version also served as the basis for the first OPC products, which were displayed at the ISA Show 1997. Consideration of further developments in Microsoft DCOM and industry requirements led to the creation of the *Data Access Specification* version 2.0 in October 1998.

Already rather early after the release of version 1.0A it could be seen that there was a need for the specification of an interface for monitoring and processing events and alarms. A working group formed to solve this problem worked out the *Alarms and Events Specification,* which was published in January 1999 as version 1.01. The *Alarms and Events Specification* defines how *Alarms and Events clients* are informed by *Alarms and Events servers* of the spontaneous occurrence of events within a process.

In addition to the acquisition of real-time data and the monitoring of events, the use of historical data offers another large field of application in automation. The work on the *Historical Data Access Specification* already began in 1997 and was completed in September 2000.

Defining and implementing security policies for use with OPC components is also of great importance. A corresponding specification has been available since September 2000 and is titled *OPC Security Specification.*

In particular from the field of industrial batchprocessing, additional requirements have been forwarded to the OPC Foundation which have led to the *OPC Batch Specification.*

During work on version 2.0 of the *Data Access Specification* and the other specifications, it emerged that there are elements common to all specifications. These elements have been combined in two specifications. The document *OPC Overview* contains explanatory aspects only, while the *OPC Common Definitions and Interfaces Specification* contains normative definitions.

With the increasing implementation of the OPC specifications in products, and their application in multiple environments, further requirements arose. With the introduction of .NET and the increasing use of Web services, OPC also faced new technological requirements. New working groups have been created by the OPC Foundation. The Data Access 3.0 working group extends the existing *Data Access Specification* with further functionality. The *OPC XML-DA Specification* defines the mapping of the *Data Access Specification* for Web services by using XML and the Simple Object Access Protocol (SOAP), which enables the use of OPC components via the Internet and on operating system platforms without DCOM. The *OPC DX Specification* defines a direct server to server communication without using a client.

The OPC Unified Architecture working group created at the end of 2003 has set itself the target of providing for global OPC-based interoperability. The definition of an interoperability platform aims primarily at improving the possibilities of combining existing OPC products (such as DA, AE and HDA servers) as well as at expanding the possibilities of use and the benefits of the different OPC specifications.

The rapid growth in the number of OPC products, from only a few in 1997 to several thousand in 2005, shows the enormous acceptance and success of this technology. OPC has succeeded not only in developing from a concept to an industrial standard within only three years, but also in becoming established in practically all areas and segments of industrial automation within the over nine years of its existence, and in playing a major role in connecting the automation level to the IT world.

1.2 OPC – an overview

OPC is the technological basis for the convenient and efficient link of automation components with control hardware and field devices. Furthermore, it provides the condition for integration of Office products and information systems on the company level such as Enterprise Resource Planning (ERP) and Manufacturing Execution Systems (MES). Process data on the field level can be presented in an Excel sheet. Status data and production data on the control level can be archived in a database without any problems via OPC or can be further processed in a production planning system.

OPC, as implemented in the currently installed product basis, is based on Microsoft's Distributed Component Object Model (DCOM). All definitions of DCOM (objects, interfaces, methods, etc.) thus also apply to OPC.

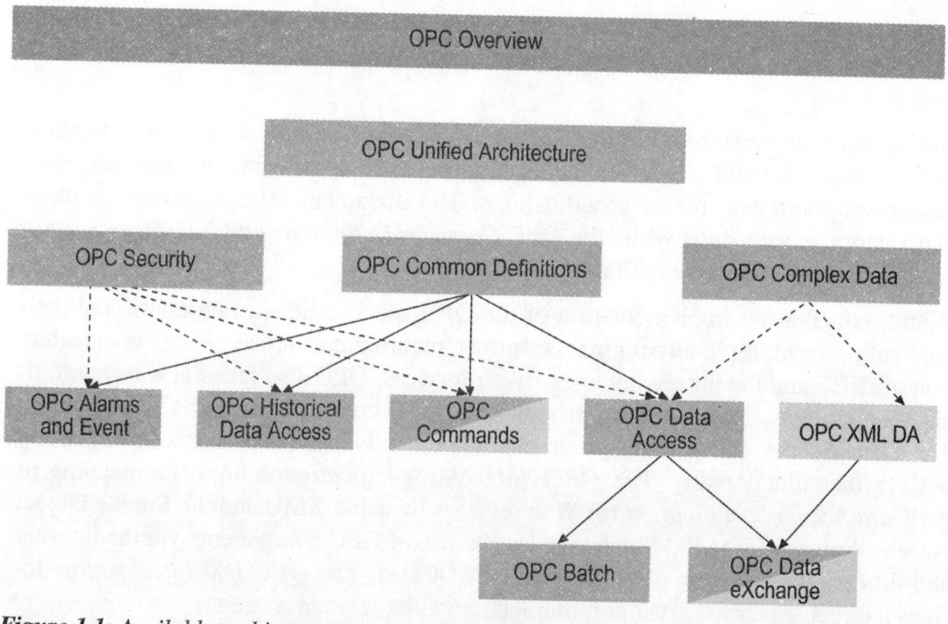

Figure 1.1: Available and in-progress specifications

The use of this technology means that the manufacturers of OPC servers and clients are not limited as to the complexity, functionality, and execution of their OPC component – as long as the appropriate definitions for the interfaces, and the methods and regulations regarding their implementation, are observed. The most diverse OPC components of different manufacturers can work together. No additional programming for adaptation of the interfaces between the components is necessary. Complex correlations, e.g. dependencies of the software component on hardware components, remain concealed behind this abstract interface. Complete components (hardware and software) can be exchanged provided the interface described in the specifications is supported.

The OPC standards are freely accessible technical specifications that define sets of standard interfaces for different fields of application in automation technology. These interfaces allow a highly efficient data exchange between software components of different manufacturers.

Figure 1.1 shows the currently available specifications or those under work and their relations to each other. These specifications concern different fields of application and are thus largely independent from one another. However, it is possible to combine them in one application.

Table 1.1 shows the contents of the different specifications and their release status.

Table 1.1: *OPC Specifications – Contents and release status (status October, 2005)*

Specification	**Contents**	**Release status**
OPC Overview	General description of the application fields of OPC specifications	Release 1.00
OPC Common Definitions and Interfaces	Definition of issues concerning a number of specifications	Release 1.10
OPC Data Access Specification	Definition of an interface for reading and writing real time data	Release 3.00
OPC Alarms and Events Specification	Definition of an interface for monitoring events	Release 1.10
OPC Historical Data Access Specification	Definition of an interface for access to historical data	Release 1.20
OPC Batch Specification	Definition of an interface for access to data required for batch processing. This specification is based on the *OPC Data Access Specification* and extends it.	Release 2.00
OPC Security Specification	Definition of an interface for setting and utilization of security policies	Release 1.00
OPC XML-DA	Integration of OPC and XML for the building of Web applications	Release 1.00
OPC Data eXchange (DX)	Communication between server and server in process	Release 1.00
OPC Commands	Definition of an interface for passing commands and supervision of their execution	Release 1.00
OPC Complex Data	Definition of possibilities for describing the structure of complex data and for accessing these data	Release 1.00
OPC Unified Architecture	Definition of an interoperability platform providing standardized access to data in DA, AE and HDA servers by means of Web services.	Draft

OPC servers can be accessed by clients written in a large number of programming languages, including C++, Visual Basic, and script languages.

What are Custom and Automation Interfaces?

Custom interfaces are used in programming languages that support the concept of function pointers (e.g. C/C++). There are, however, programming languages (e.g. Visual Basic or Visual Basic for Applications) that do not use any function pointers. To enable clients written in those languages to access DCOM objects, automation interfaces have been introduced. Automation interfaces use a standard interface where methods are no longer called by function pointers but by their name.

An automation object implements properties. These represent rather static settings of an object. They can be set or retrieved. The functionality of the object can be used by the client by means of methods. The client is informed of special occurrences by events.

All information on the automation objects of a component is included in the Type-Library. This library can be loaded in the development environment and enables easy access to information on methods, events, and properties.

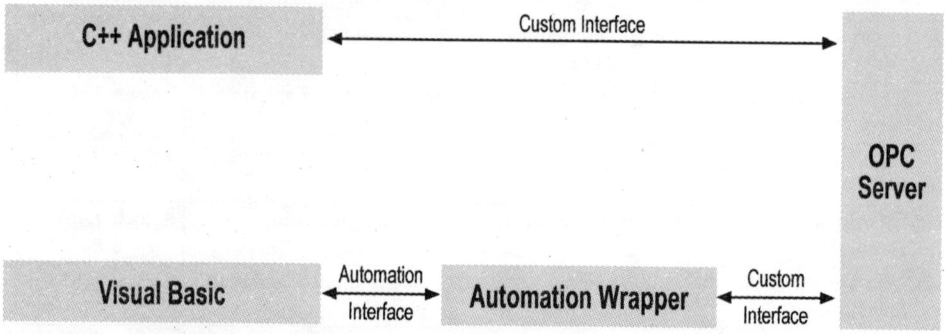

***Figure 1.2:** Access to OPC servers using Custom and Automation interfaces*

The automation interfaces defined in the OPC specifications are combined in an Automation Wrapper. The Automation Wrapper is a DLL that forwards calls from the methods at the automation interface to the custom interface. **Figure 1.2** shows the various possibilities of access to the OPC server. The source code of the wrapper is made available by the OPC Foundation. However, it is more likely to be seen as sample code. If a manufacturer wants to provide the wrapper as component of its product, it has to create its suitable version, which will be subject to the usual tests. The different objects also offer all the methods available at the corresponding custom interfaces of the objects.

Table 1.2 below describes the advantages of using OPC. All those involved in the net product chain are considered, from development up to the application of OPC technology:

Hardware manufacturers

Manufacturers of devices that provide or need data (e.g. PLC, barcode reader, measuring instruments) or manufacturers of PC plug-in cards that enable access to such devices (e.g. field bus boards, measuring systems).

Software manufacturers

Manufacturers of applications for the acquisition, visualization, or processing of data of the above mentioned devices and for sending data to those devices.

System integrator

Prepares complete solutions based on the products available on the market.

End user

Uses solutions based on OPC products.

***Table 1.2:** The advantages of OPC*

Hardware manufacturer	The product can be used by all OPC-compatible systems in the market and is not limited to an individual system for which a corresponding solution (i.e. specific drivers) must be developed. Due to the existence of standardized interfaces and the interoperability related to them, there is no need to become familiar with the specific requirements of other systems. The Time-to-Market for new device generations is significantly reduced as only one OPC server has to be updated, not a large number of drivers or DDE servers. The effort needed for support is also reduced as only one product has to be supported.
Software manufacturer	The product can be used with all devices and communication protocols on the market that make available an OPC interface. The manufacturer no longer has to develop corresponding solutions (specific drivers). Due to the existence of standardized interfaces and the interoperability related to them, there is no need to become familiar with the specifications of other devices and communication protocols. The time needed for support is considerably reduced as many products to be supported (product specific drivers up to now) no longer exist.
System integrator	Flexibility in the choice of product is considerably increased, as is, consequently, the number of projects that can be processed. The time needed for integration and training is considerably reduced, as OPC provides a standardized interface which remains the same for all products.
End user	OPC provides additional flexibility (distribution of components, use of new technologies, choice between products, etc.) during the design of the overall system as products of various manufacturers can be combined. The increasing variety of products available on the market should, in the future, lead to a noticeable reduction of costs with increasing quality and user comfort. By relying on an acceptable standard in this sector, investments will be better protected.

1.3 Comparison and assessment of technologies

Table 1.3: *Comparison of technology*

	Product-specific drivers	DDE	OPC
Standardization	No	No	Yes
Interoperability	No	Limited either to standard DDE or to the respective dialects (e.g. FastDDE, AdvancedDDE).	Yes
Performance	Very high performance as the solutions are optimized as to the specific application.	DDE is based on the exchange of messages. This is slow. DDE dialects improve efficiency by putting more data into one message.	Very high performance. For concrete figures, see chapter 4.5.1.
Remote access	No	Yes (through NetDDE)	Yes (Distributed COM), Web services
Interactions	Generally not available.	No	Yes. OPC defines a large number of methods, which can be used for exchanging additional information between client and server.
Complexity	Many manufacturers provide toolkits with which drivers can be created quickly for corresponding systems. A new system requires additional time to learn.	Toolkits are available from manufacturers. As the variety of DDE shapings is not very high, familiarization with them is rarely necessary. Programming of DDE applications without toolkits is very difficult.	Learning how to program DCOM applications and Web services is time-consuming. There are toolkits that help minimize this time.
Market share	The market share is reduced with increasing speed.	DDE servers will disappear in future.	The share of OPC solutions will dominate.
Availability on other platforms	There are only a few systems available for different platforms.	DDE is only available for Windows operating systems.	The OPC components available today are limited to systems where DCOM is available. These are primarily Windows systems. The *OPC XML-DA Specification* additionally allows the use of OPC components on operating systems without DCOM.

Table 1.3 compares OPC, DDE, and product-specific drivers with regard to the following criteria:

Standardization

Are there any internationally approved definitions for the use of this technology in an automation environment?

Interoperability

Can solutions from different manufacturers work with each other?

Performance

Efficiency with regard to the time needed for data transfer is assessed.

Remote access

Is it possible to create a distributed system?

Interactions

Does functionality go beyond the pure exchange of process data? This concerns e.g. querying of structural information at run time.

Complexity

What kind of knowledge is required to create a corresponding product?

Market share

What is the market share of the technology today and what will be the future development?

Availability on other platforms

Is it possible to use the technology on non-PC platforms and non-Microsoft operating systems?

1.4 Structure and tasks of the OPC Foundation

An important prerequisite for the success of a standardization initiative is an authority coordinating the interests of the members involved. The task of this authority is to protect the common objective from the political interests of individuals. Specification work has to be initiated and marked with clear mission statements in order to avoid a proliferation of variants and derivatives. Furthermore, public relations have to supply the market with information and support the common standard.

The OPC Foundation was founded in 1996 as an independent non-profit organization with the aim of further developing and supporting the new OPC standard. The OPC Foundation rapidly developed into an internationally recognized organization. In the over nine years since its foundation in 1996 and in view of its large and still-growing membership, the OPC Foundation succeeded in maintaining a uniform OPC standard worldwide, which has not drifted apart into different, incompatible local variants and derivatives.

As a non-profit organization, the OPC Foundation is dependent on specification work done by "volunteers" – members who are partly released by their employers for this purpose. On the one hand, this has the big advantage that a specification can be made available free of charge after completion and thus can be implemented quickly and repeatedly without paying high costs. On the other hand, important tasks and time pressure in a company may limit the commitment of volunteers and thus put the rapid progress of the specification work at risk. Besides specification development, other tasks are distributed to different offices and persons within the OPC Foundation.

The structure of the OPC Foundation's organization is shown in **Figure 1.3**.

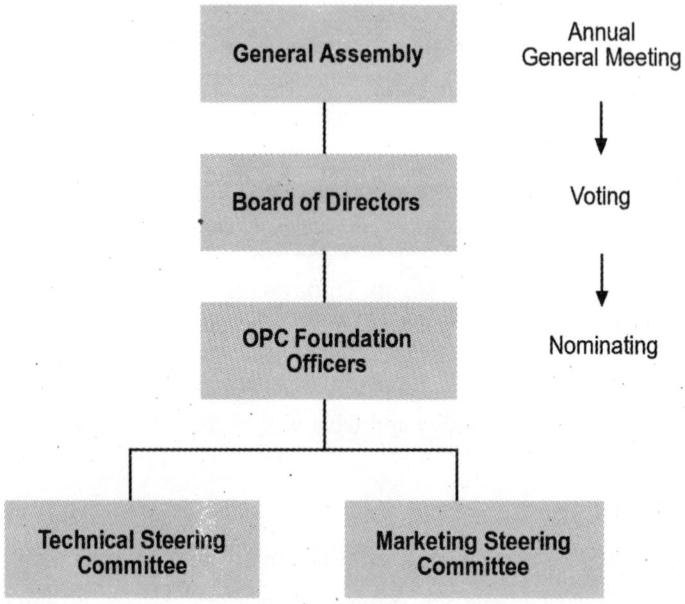

Figure 1.3: *Structure of the OPC Foundation*

The Board of Directors is the Foundation's decision-making body. It is elected once a year at the General Assembly by the members entitled to vote. In 2004/2005, the board consists of representatives of the following companies:

Emerson Process Management, Honeywell, Iconics, Rockwell Software, Siemens AG, Yokogawa.

Once a year, the Board of Directors appoints the persons who safeguard the interests of the Foundation between the general meetings.

For 2004/2005, the following officers have been nominated:

President:	Tom Burke	OPC Foundation
Vice President:	Reinhold Achatz	Siemens AG
Treasurer:	David Eisner	Honeywell
Secretary:	Michael Bryant	OPC Foundation

Specification work is in principle centrally coordinated by the Technical Steering Committee (TSC). The TSC consists of representatives of the same companies as the Board of Directors plus the chairmen of the working groups. The head of the TSC is Jim Luth of Iconics. The TSC establishes working groups for several target projects.

The Marketing Steering Committee is managed by Russ Agrusa of Iconics.

There are two kinds of OPC Foundation members: the OPC technology users and the OPC technology providers. The latter are further categorized as profit and non-profit organizations.

The annual membership fees for technology users and non-profit organizations are independent of their size. The annual membership fees for the technology providers depend on their annual turnover. **Figure 1.4** shows the division of the firms by category and turnover.

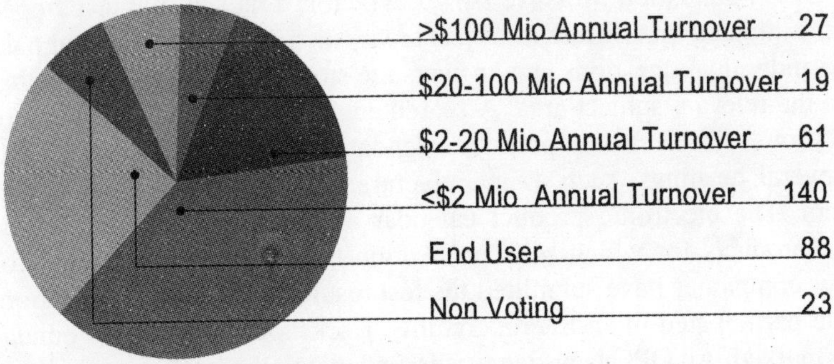

Figure 1.4: *Division of OPC-Foundation members by company category (May 2005)*

A General Assembly is held once a year. The representatives of the member companies can participate "virtually" in the form of a Web conference. Sound is transmitted via dialup connection to a telephone server while the visual presentation is shown at a web site to which the member companies can log on with a password.

The OPC Foundation in October 2005 had 362 members worldwide. The member firms from Europe and the Far East are organized in their own OPC subcommittees. **Figure 1.5** shows the regional division of the member companies.

OPC Europe, OPC Japan and OPC China represent the members of these regions and in addition support OPC by providing information to members in the regional languages. They conduct their own events, such as participation in fairs or interoperabil-

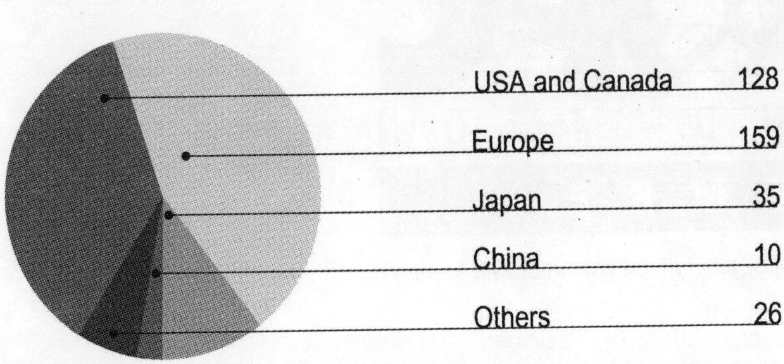

Figure 1.5: *Division of OPC Foundation members by region (May 2005)*

ity workshops. They are coordinated with the central organization via the Board of Directors and also play their part in the annual general Assembly.

The OPC Foundation has an Internet Web site. At www.opcfoundation.org, visitors can find information on the organization, its members, the working groups, and current events. Furthermore, visitors can download released specifications and technical reports in the form of "white papers." A computer-aided process for publishing Web updates allows making innovations available very quickly. The member companies can announce new products, edit their company details, update their product information and publish press releases. Questions to the OPC Foundation are quickly answered by the OPC Foundation Message Board, which is a database-assisted open forum. The forum is divided into different subject areas, such as specifications or base technologies. The individual questions and answers are stored and displayed together with keywords in the relevant subject area. A search engine enables the fast query of all answered questions. An electronic product calendar lets visitors search for OPC subjects under several headings, such as manufacturer, client, server, development tool, and seminars. The electronic product calendar additionally contains the test records of all the products for which self-certification has been performed and for which the member companies have submitted the test results to the OPC Foundation. Products that have participated in an Interoperability Workshop of the OPC Foundation are specially marked. An OPC Newsletter is issued two to three times annually at the OPC Foundation's web site. It contains the latest news about the organization, an event calendar as well as user reports.

From the central Web site of the OPC Foundation, the links www.opceurope.org, www.opcjapan.org and www.opcchina.org lead to the sites of the European, Japanese and Chinese subcommittees.

The OPC Foundation provides a membership application form on its web site. Members of the OPC Foundation may use the logo shown in **Figure 1.6** for public relation activities.

Figure 1.6: Logo to be used by OPC Foundation Members

The benefits and rights of membership are shown in the **Table 1.4**:

Table 1.4: *Types of membership in the OPC Foundation*

	Corporate Member	End-User Member	Non-voting Member
Voting Rights	Yes	Yes	No
Participation in events of the OPC Foundation			
Participation in Trade Shows & Demonstrations	Yes[2]	May be listed in advertisement and publicity	May be listed in advertisement and publicity
Participation in the OPC General Assembly	Yes	Yes	Yes
Participation in Microsoft/OPC NDA Technical Updates	Yes	Yes	Yes
Participation in Interoperability Workshops	Yes	No	No
Availability of OPC Specifications	Early access to Specification and sample code	Early access to Specification and sample code	Access to completed Specification
Eligibility for Committees			
Technical Steering Committee	Yes (must be appointed by the Chair of the TSC)	No	No
Technical/Marketing SubCommittee	Yes	Yes	No
End-User Advisory Council	n/a	Yes	n/a
Services			
Listed in OPC Membership Directory	Yes	Yes Separate Mention	Yes
Company Profile listed in Product Vendor Directory	Yes	n/a	No
Products listed in Product Vendor Directory	Yes	n/a	No
Use of OPC Member Logo in Advertising	Standard	Standard	Standard
Use of OPC Compliance Logo for Certified Products	Yes	n/a	No

Specification Work

Specification work counts among the tasks of the OPC Foundation with the greatest importance. The specification process has to be clearly defined, progress has to be monitored, and the results have to be released.

2 Subject to cost sharing

First of all, the Board of Directors defines the specification issue in the form of a mission statement and appoints the chairman of the working group. The chairman addresses the member companies of the OPC Foundation and asks them to cooperate. Interested companies then appoint members. The working group meets several times and prepares the specification. Creation of sample code, which proves the principal implementation possibility and use of the new specification, is effected in parallel. The specification and the sample code are then passed to the Technical Steering Committee for approval. If approved, the specification and code are submitted to the Board of Directors for release. Otherwise they are returned to the working group for further processing.

1.5 Collaboration with other organizations

The growing importance of OPC, the use of OPC products in more and more new application areas, and the new technological requirements, such as platform independence and Internet capability, has created points of contact with other organizations. When entering new application areas, the OPC Foundation does not aim at superseding existing standards of other organizations or reinventing their activities and specifications. The objective is to combine the strengths of the relevant organizations in such a way that OPC adds "how" data are transported while the other organizations deal with "what" is to be transported, i.e. what data and what information models on which these data are based. This is a further significant potential of OPC: The newly created interoperability provides for a more widespread use of the existing standards of other organizations and their products. Collaborations already exist in the plant maintenance and semiconductor areas as well as with fieldbus organizations in the area of device description, and in building automation and ISA. The points of contact between OPC and the other standards vary significantly. With regard to plant maintenance, the OPC Foundation has founded an OpenO&M initiative together with the Mimosa organization. Mimosa has defined a standard for the operations and maintenance (O&M) of plants and factories. OpenO&M aims at making O&M applications easy to integrate into a multi-vendor environment according to the Mimosa model via XML-based OPC. For the different fieldbus organizations, *OPC DX* bridges incompatible communication protocols and allows horizontal interoperability. The *Complex Data Specification* and the new *OPC Unified Architecture Specification* enable using the electronic device description language EDDL for describing the data structures of FOUNDATION Fieldbus, HART or PROFIBUS objects. In building automation, OPC provides an open interface for the bus systems used in this industry, such as Bacnet, LON etc. The "Open Interfaces" working group of the German ZVEI (Zentralverband Elektrotechnik- und Elektronikindustrie e.V.) has developed a concept covering fire alarm, fire fighting and burglar alarm systems as well as video monitoring and access control systems, and various other systems. In this concept, OPC is defined as the interface from these systems to the alarm management systems. The concept also stipulates the application of the OPC Security Interfaces definitions laid down by the OPC Foundation in the *OPC Security Specification*. The "Arbeitskreis Maschinen- und Elektrotechnik staatlicher und kommunaler Verwaltungen (AMEV)" working group, which is in charge of the machinery and the electrical installations in municipal and state administrations in Germany, has included OPC as the interface standard into its building automation recommendations. These examples as well as many others demonstrate the great acceptance of OPC and the tremendous opportunities that open up for other organizations through the adoption of OPC as the interface standard.

The next chapter describes the existing specifications and those under preparation:

- *OPC Overview* and *OPC Common Definitions and Interfaces* (Section 2.2)
- *OPC Data Access V2.05* (Section 2.3)
- *OPC XML-DA* (Section 2.4)
- *OPC Data eXchange* (Section 2.5)
- *OPC Complex Data* (Section 2.6)
- *OPC Command Execution Interface* (Section 2.7)
- *OPC Data Access V3.0* (Section 2.8)
- *OPC Alarms and Events* (Section 2.9)
- *OPC Historical Data Access* (Section 2.10)
- *OPC Batch* (Section 2.11)
- *OPC Security* (Section 2.12)
- *OPC Compliance Test* (Section 2.13)
- *OPC Unified Architecture* (Section 2.14)

2 Fundamentals

2.1 Introduction

This chapter deals with the explanation of the OPC specifications. It is not intended to replace study of the specifications, its purpose is to present the facts – which are often dealt with in a prosaic manner – on the clearer basis of application possibilities. Therefore, we largely do not discuss methods, parameters, or interfaces in detail. Instead, functionality is explained by means of scenarios.

In the following sections, all OPC specifications released or in preparation are dealt with. Reference is always made to the release status of the specification under discussion.

Every specification is destined for a specific field of application, and is based on a model that corresponds to this field of application. In this chapter, the presentation of a specification begins with these two items. These are followed by a description of the specification objects and their possible uses. As some specifications are longer than others, the individual sections describing them, too, differ in length.

This chapter concludes with a section describing the use of the OPC specifications.

Note: OPC specifications cannot be explained without using DCOM terms. The basic terms required are explained in sections 1.1 and 1.3. A detailed explanation of DCOM is available in section 3.1. OPC specifications today use Web services as an additional technological basis. This technology is described in section 3.4.

2.2 OPC Overview and OPC Common Definitions and Interfaces

2.2.1 Field of application and model

The explanations in this section relate to version 1.0 of the *OPC Overview* and version 1.10 of the *OPC Common Definitions and Interfaces* specification.

During the specification work, it became clear that there are definitions and general statements common and relevant to all specifications. These definitions and statements were combined in the specifications *OPC Overview* and *OPC Common Definitions and Interfaces*.

These specifications were created following the work on version 2.0 of the *Data Access Specification* and the first versions of the *Alarms and Events* and *Historical*

Data Access specifications. These specifications define object models, interfaces, methods, parameters, and concepts for different fields of application.

The *OPC Overview* contains informative statements on the following subjects:

- Field of application of OPC
- Basic technology for OPC
- Specifications available

Due to its informative character, this specification is not discussed any further.

The specification *OPC Common Definitions and Interfaces* contains, as its name indicates, general definitions and interface descriptions.

These include:

- the interface *IOPCCommon*
- the interface *IOPCShutdown* at the client side
- Installation definitions and the description of registry settings
- the *OPCServerBrowser* and
- definitions for specific ID subranges of properties

2.2.2 Specification object and use

This section describes the functionality defined in the *OPC Common Definitions and Interfaces Specification.*

Installation and Registration requirements

OPC clients and servers share some components in common. These are:

- Proxy/Stub Files, which provide for marshalling and de-marshalling of the interface pointers and method parameters. An explanation of this concept can be found in section 3.1.
- OPCServerBrowser. The component described below offers an interface that enables OPC clients to inform themselves of OPC servers on remote computers.
- Automation Wrapper. Besides the opportunity to access the functionality of the server by the custom interface, this access is also possible via automation interfaces.

How can it be guaranteed that these components are available on the computer as long as there are OPC clients and servers available?

The prerequisites have to be created during installation and un-installation.

To do this, nothing has to be newly invented; the concept of shared files (shared libraries), already supported in the Windows operating system, is used. A counter in the registry allocated to each of the components is increased by one for every installation and decreased by one every time the component is uninstalled. If the counter reaches the value zero, i.e. there are no OPC products on the computer any longer, the corresponding component is deleted. Current installation programs support this procedure.

An additional definition concerns the question of when to install the OPCServer-Browser – during installation of the client, or the server? The specification defines that the manufacturer of the server is responsible for this.

Since 2003, the OPC Foundation has provided an installation for the components listed above, which are called core components. The installation package is available free of charge at the OPC Foundation web site.

The client needs the CLSID of the server to use the DCOM runtime environment for instantiation of a DCOM object (This process is described in section 3.1.). Where does the CLSID come from? One possibility would be for the operator to take the CLSID out of the documentation and to pass it to the client. However, that is not practical. For this reason, another way is provided in the specification.

Every OPC server has to make at least three entries in the registry:

1. The ProgramIdentifier (ProgID) contains a readable string that describes the component. The structure of the string is predefined: <Manufacturer>.<Server name>. The contents can be determined by the manufacturer: Softing.OPC_SMPL_SRV_ENGINE. The string may not contain any blank spaces.
2. The ClassID (CLSID) is a 128-bit value which uniquely identifies a server. This value is allocated to exactly one server, and this server is always linked to this CLSID. If structural changes are made in the server (adding of new objects and interfaces), a new CLSID has to be used. The GUID can be generated with the *guidgen* program, which is part of Visual Studio.
3. The ApplicationID (AppID) contains another readable description of the server. The structure and contents of the AppID are not predefined.

In the scope of the *OPC Data Access Specification* 1.0A, it has been determined that the entry of the ProgID of a server in the registry has to have the subkey "OPC". This means that there is a difference between the entry for an OPC server and that for other DCOM components. A subkey of the ProgID contains the CLSID.

As there are several specifications and thus several categories of OPC products available, this "OPC" subkey alone is not sufficient. For example, a *Data Access Client* wants to communicate with a *Data Access Server* and not with an *Alarms and Events Server*. That is why distinctions are made between the various specifications by means of CategoryID (CATID). The CATID is situated in a subkey of the CLSID. CATIDs are defined in the various specifications. **Table 2.1** shows the current values.

Table 2.1: *OPC specifications and their CATIDs*

Data Access Specification 1.0A	{63D5F430-CFE4-11D1-B2C8-0060083BA1FB}
Data Access Specification 2.0	{63D5F432-CFE4-11D1-B2C8-0060083BA1FB}
Alarms and Events Specification 1.0	{58E13251-AC87-11D1-84D5-00608CB8A7E9}
OPC Batch Specification 1.0	{A8080DA0-E23E-11D2-AFA7-00C04F539421}
OPC HDA Specification 1.0	{7DE5B060-E089-11D2-A5E6-000086339399}
OPC Batch Server Version 2.0	{843DE67B-B0C9-11d4-A0B7-000102A980B1}

The entry for the AppID contains subkeys that are also mentioned in section 3.5, "Security."

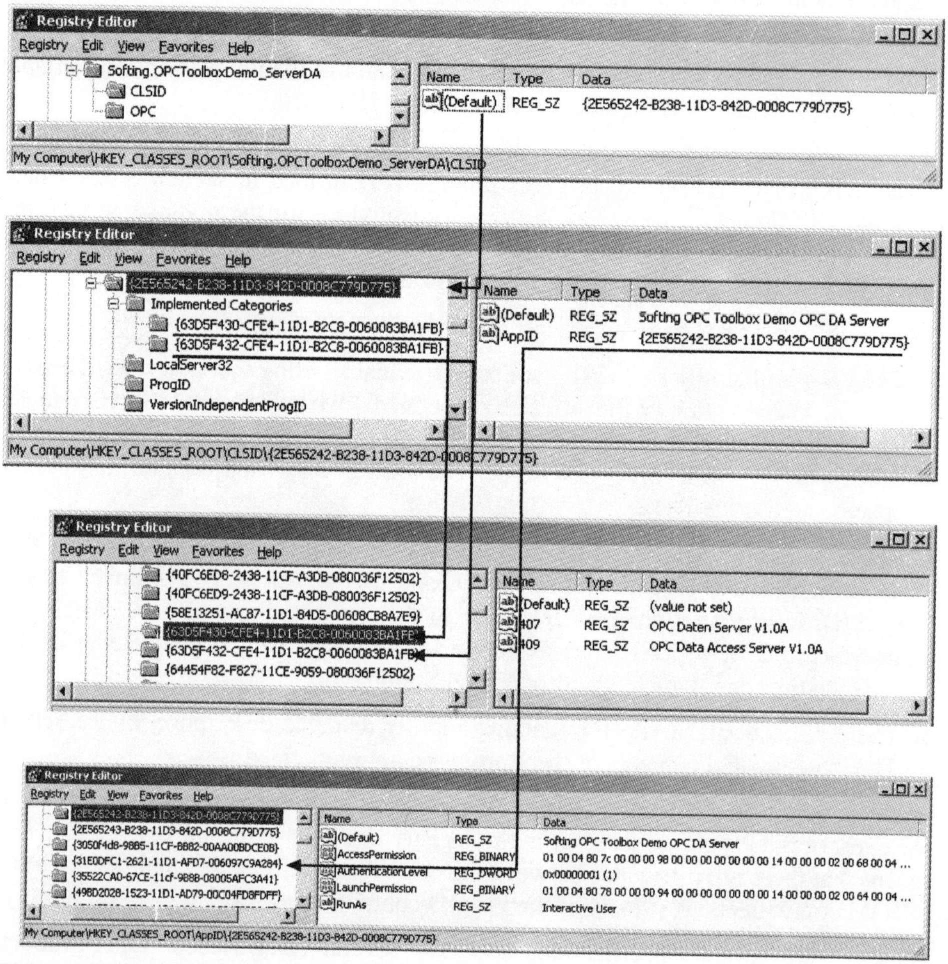

Figure 2.1: OPC specific registry entries and their dependencies

Figure 2.1 shows the different entries in the registry and their relation to each other. The entry for the ProgID is found by searching for the subkey "OPC." The subkey CLSID contains the ClassID. The ClassID is still available at another place in the registry. This ClassID, in turn, also has subkeys, e.g. AppID and CATID.

OPCServerBrowser

The information needed by a client to launch a server is available in the registry, as explained above. Browsing the registry is a programming task for the client, which can be easily performed provided a local resource is available. Should browsing of the registry be carried out on a remote computer, however, problems may arise due to access restrictions. These problems can be remedied with the OPCServerBrowser made available by the OPC Foundation. This component is a DCOM server with a

CLSID defined in the specification. Unlike the CLSID of other servers, it can be firmly determined during programming of the client. The client is thus not forced to search for it in the registry. To get the CLSID of OPC servers, the client uses the functionality of the OPCServerBrowser.

The OPCServerBrowser offers the following methods at the interface *IOPCServerList*:

- *EnumClassesOfCategory*: When calling this method, the client transfers one (or several) CATID(s) and gets back one or more lists of CLSIDs of the servers that implement the corresponding specifications. In a real application, the client would transfer {63D5F432-CFE4-11D1-B2C8-0060083BA1IB} as a CATID for Data Access Servers implementing the functionality of version 2.0. The ServerBrowser would then return the CLSIDs of the corresponding servers located on the computer, if any.
- *GetClassDetails*: The client transfers the CLSID and gets back the ProgID and the description of the server. This value can be found by the OPCServerBrowser in the AppID entry. With the method call EnumClassesOfCategory, the client has received the CLSID. Offering this information for selecting a specific server is not very user-friendly. By calling the function GetClassDetails, the client requests the corresponding information.
- *CLSIDFromProgID*: The client transfers the ProgID and gets back the CLSID. This method is used if the ProgId is already known (e.g. in the Automation Wrapper for Data Access), but the program still needs the CLSID for starting the server.

The OPC client will use the functionality of the OPCServerBrowser both on the remote computer and on the local computer.

The IOPCServerList2 interface was introduced with version 1.10 of the specification. This interface solves a problem that occurred due to the close link between the functionality of the IOPCServerList and the Internet Explorer 5. An interface used in the IOPCServerList was not supported by the more recent versions of the Internet Explorer so that the IOPCServerList could not be used any longer.

Possibilities of adaptation of an OPC server to desired languages

Some methods at interfaces defined in the OPC specifications can transfer information on the success or failure of their execution to the client in the form of a numerical value. More meaningful, however, is textual information. It would be even better to get this information in a desired language – to localize the error text. For this purpose, methods are available at the interface *IOPCCommon*. This localization of error information was the original intention for the definition of this functionality. The scope of the language identifier has been enlarged in the meantime and can be used for almost all language-dependent parameters.[3] An exception (which is also defined in the specification) is the description of the properties (see section 2.3.1). This description is always in English.

3 Besides English and German, the supplied sample server also supports Swedish. The sample client can query which languages are supported and specify the required language.

By using *QueryAvailableLocaleIDs*, the client can ask the server for the languages supported. Examples for LocaleIDs are 0x0800 for system default and 0x0409 for U.S. English. High-quality servers will offer a wide variety of language support. The client can set a standard language with *SetLocaleId* which will be used implicitly with all subsequent queries of *GetErrorString*. *GetErrorString* is used to get a textual description of the error. By using *GetLocaleId*, the client can query for the current default language.

The client can transfer information to the server by using the method *SetClientName*. The specification recommends that this includes:

– the name of the executable file which contains the client (e.g. "application.exe")

– the name of the computer on which the client is running (e.g. "Visualization").

All methods are specific to one client. If multiple clients access the server at the same time, the settings are made separately for each client.

Transfer of shutdown request from server to client

What happens when an OPC server has to be shut down while it is being used by an OPC client? In this case, the server informs the client of this event. The server calls a method at the corresponding callback interface of the client and informs the client of the forthcoming shutdown. As a method parameter, the server can pass a text string the semantics of which are not defined. The behavior of the client and the server after calling the method is not defined. This subject of the specification applies to a foreseeable interruption of the server connection. Version 3.0 of the *Data Access Specification* defines the behavior in the event of an unforeseeable interruption of the connection between client and server.

2.3 OPC Data Access Specification

2.3.1 Field of application and model

The explanations in this section refer to version 2.05 of the specification. Concepts of previous specifications – those which have been complemented by other concepts but which are still valid – are mentioned as well.[4]

The OPC *Data Access Specification* 2.05 today is implemented in a vast majority of the products. It will therefore be explained in greater detail in this section. The *OPC Data Access Specification* 3.0, in contrast, is only used by few products today (2005). For this reason, only the additions and amendments to this specification as well as their consequences will be described later in this chapter.

4 Due to definitions in DCOM, declarations for interfaces cannot be modified. If a change in the existing funktionality is required (e.g. an extension of funktion parameters). new interfaces have to be introduced. The old interfaces, however, have to be further implemented to enable interoperability between components of different edition status. This is why the definitions of specification 1.0A are also included in specification 2.05. Therefore, no explicit reference is made to specification 1.0A here.

The *Data Access Specification* is the oldest of all the OPC specifications. In its first version, it was simply called the *OPC Specification*. It defines an interface between client and server programs to access process data. *Data Access Servers* provide one or several *Data Access Clients* with transparent access to different data sources (e.g. temperature sensors) and data sinks (e.g. controllers). These data sources and data sinks can be located on I/O cards directly plugged in into the PC; they can also be located on devices such as controllers and input/ output modules. connected by means of serial links or via field buses. Of course, it is possible for a *Data Access Client* to have access to several *Data Access Servers* at the same time.

Data Access Clients can be very simple (e.g. Excel sheet) or complex (e.g. Visual Basic Visualization). *Data Access Clients* can also be components of larger programs, e.g. HMI or SCADA systems.

Data Access Servers can be simple programs, for example one which provides access to the registers of a PLC via a serial communication link. More complex programs are also possible, however, enabling access to a large number of variables in a variety of devices via extensive communication mechanisms. *Data Access Servers* can also be components of larger programs (e.g. PC based Control) and can make available the data of these programs. The specification does not contain any definitions or limits for these implementation aspects.

The *Data Access Specification* defines two different concepts which a *Data Access Server* can implement and a *Data Access Client* can use – the namespace and the OPC object hierarchy.

The namespace contains all the data sources and data sinks made available by a server. The namespace can be a tree structure of any depth (hierarchical namespace). It can also be flat, however (flat namespace). In a flat namespace, all the leaves are on a single level. There are no nodes.In a hierarchical namespace, nodes can be used as structuring means; they can for example represent devices in an installation. Leaves are available at the nodes, representing data sinks and data sources. These can be setpoints and measured variables of a device. The leaves are also called *Items*.

In an application, not only are data sources and data sinks important, but also the information on these variables and devices. Copying this information on items unnecessarily expands the namespace. Therefore, attributes, so called *Properties,* are allocated to nodes and leaves. The name of the manufacturer of a device, for example, can be mapped to a *Property.*

In addition to properties, *AccessPaths* can also be available at an item. They can be used to describe the communication path between the *Data Access Server* and the device (e.g. COM1, 9600kBit/s) in more detail. The *AccessPath* is optional and can be omitted.

A *Data Access Client* can create several OPC objects in a *Data Access Server* to define its view of the process. The *OPCServer Object* is the top level object in the hierarchy of the OPC objects. *OPCGroup Objects* form the next level. The last level is formed by *OPCItem Objects*. OPCItem Objects represent leaves or properties of leaves and nodes of the namespace, i.e. the parameters that the client is interested in.

OPCGroup Objects are used to structure OPCItem objects. This can be done for logical aspects (e.g. all variables in one process section) or dynamic aspects (e.g. process variables with the same or a similar time behavior), as the user likes. All the OPCGroup objects are managed by the OPCServer object. There can be several of these object hierarchies available within one OPC Server component if several *Data Access Clients* have access to this component. For every client, a specific hierarchy exists which exactly reflects its requirements.

The number of hierarchy levels for the OPC objects in the server is fixed. **Figure 2.2** shows an object hierarchy and the relations to the items in the namespace. The namespace only exists once and is the same for all clients. The client shown in **Figure 2.2** has created two OPCGroup objects and, for each OPCGroup object, two OPCItem objects.

In the forming of designations for the *Data Access Specification,* unfortunately, the necessary care was not taken everywhere. In oral communications, the terms of the object OPCServer and the component OPC server can often be mixed up. In this book, the term OPCServer is only used for the Top-Level object within a *Data Access Server.* The two-word phrase OPC server is used for all the servers that can be created based on OPC specifications.

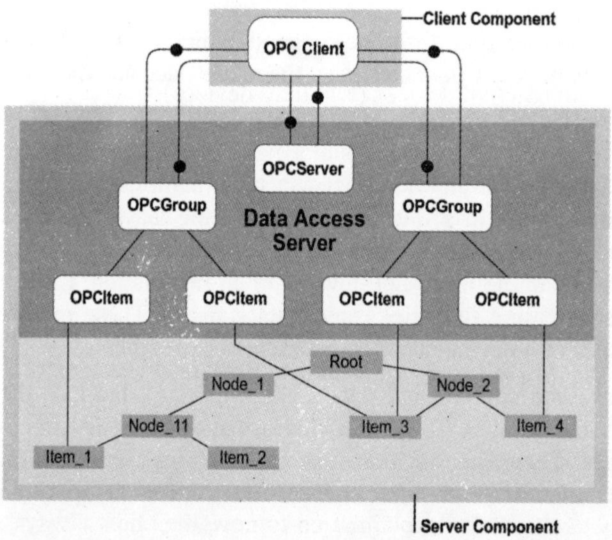

Figure 2.2: Data Access Server – Namespace and Object hierarchy

2.3.2 Specification object and use

The first two specification objects – Object hierarchy and namespace – have been mentioned in the previous section.

Only two objects from the object hierarchy are "real" DCOM objects – OPCServer and OPCGroup. For these two objects, interfaces, methods, and parameters have been defined. The object OPCItem has no custom interfaces. Why? In real applications, several values are read and written at the same time. It is more efficient to access several OPCItem objects with a single call. This is done via methods at interfaces of the OPCGroup object. For this reason, OPCItem objects do not have any custom interfaces and their implementation can be kept simple. This model does not exclude read or write access to a single OPCItem object. Automation interfaces have been defined for all three groups of objects.

Besides these provisions, there are definitions of different types of data exchange between *Data Access Server* and *Data Access Client*. They are shown in **Table 2.2** below. The values can be taken from the server's internal cache of the process data (Cache) or directly from the data source (Device).

Table 2.2: *Types of data exchange between Data Access Servers and Clients*

	Cache	**Device**
synchronous Read	x	x
asynchronous Read	x[5]	x
Refresh	x	x
Subscription		

Three basic possibilities can be distinguished for how data access is carried out by *Data Access Servers*:

– The values are available in the PC, e.g. in another application.

– The values have to be called from external devices (e.g. field devices).

– The values are sent by the external devices without being asked.

By using synchronous or asynchronous calls, the client can adapt optimally to this situation. With synchronous reading, the client calls the method and waits for the return value. Synchronous calls are only to be used if data access takes place quickly (e.g. values are available in the PC) as otherwise the performance of the client could be affected by the handling of outstanding calls. With asynchronous reading, the client calls the method of the server and immediately gets feedback. After a certain interval, which depends on the kind of data access, the client gets the desired value. Asynchronous reading is to be used if data recording through the servers takes longer (values have to be requested).

With a refresh, the client reads all active OPCItem objects of an active OPCGroup object. For the three types of data exchanges explained up to now, the client always takes initiative.

This is not always efficient, as the client will call values regardless of possible changes. It is important, however, only to transfer a value from the server to the client if the value itself has changed. For this reason, a fourth type of data exchange exists.

In this manner, the server reads the values in cycles determined by the set update rate and passes them to the client if a change in the values or the status has taken place.

5 Reading of the cache has been defined in Data Access Specification 1.0A. Data Access Specification 2.0 defines a new interface which supports asynchronous reading only for the device.

The client is able to adapt the update rate and the sensitivity to concrete monitoring requirements.

Besides the reading functionality, the procedure of synchronous and asynchronous writing was also determined. Writing actions are always effected directly in the device. Writing to a cache and automatically transferring the values at a later date is not possible.

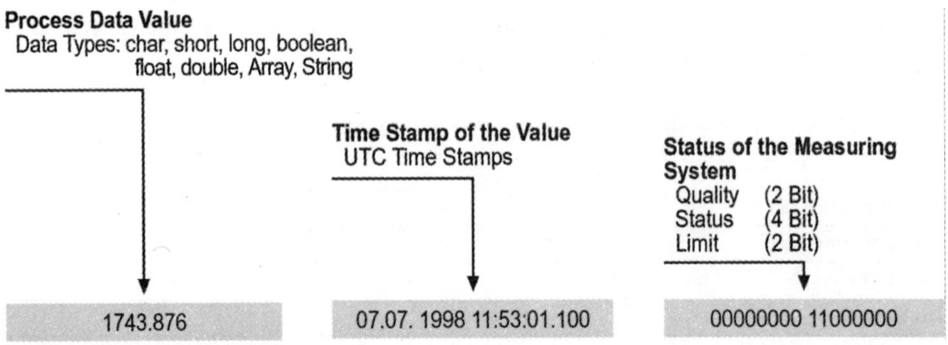

Figure 2.3: *Data format of the Data Access Specification*

Furthermore, the format in which data are exchanged between the server and the client has been established. This format is shown in **Figure 2.3**. It consists of three components:

- The actual data. OPC allows interoperability between client and server products from different manufacturers. The servers provide transparent access to data in different application areas. The clients make data available to other applications. As different type systems are used, a standard type system must be supported to ensure interoperability. OPC uses the DCOM VARIANT type system. To exchange data between the client and the server (and vice versa), the information is mapped to DCOM VARIANT data types. The server and the client are responsible for converting DCOM data types into other data types, and vice versa.
- The time stamp. There are devices that supply the time stamp with the value itself. In this case, it is used; otherwise, it is generated in the server. The time stamp is 8 bytes long and indicates the time from 01.01.1601 in ticks of 100 ns. The time stamp is a standard structure for the representation of Universal Time Coordinated (UTC) in Windows systems.
- The status information. This bit field of 2 bytes comprises status information for the data transmitted. For the time being, only the use of the less significant byte is specified. Two bits are used to describe the quality of the value. The value can be Good, Bad, or Uncertain. The server will mark the value as "Bad" if there is, for example, no link to the device. It will mark it as "Uncertain" if, for example, there is a link to the device but the value makes no sense because, for instance, there is a deviation from the measuring range. Four status bits give a more detailed description of the current status. An example of this is the value "Not Connected" as a more precise explanation of the status "Bad". "Last usable Value" is a possible substatus of the status "Uncertain". The possible bit values, their causes as well as a

description are defined in the specification. The last two bits can contain additional diagnostic information in case of a sensor failure. This means that a value can be marked as not plausible.

The specifications of how – and in which format – data are exchanged between a *Data Access Client* and a *Data Access Server* allow for the concealment of differences between servers with regard to how they access data (e.g. from a database or via a serial protocol).

Using the example of a server that records data via a field bus, **Figure 2.4** shows the basic structure of the implementation of a *Data Access Server*. Distinctions can be made between an application-independent part and an application specific part. The first contains the implementation of the objects OPCServer and OPCGroup and parts of the OPCItem objects. A further part of the OPCItem objects is application-specific, e.g. for a field bus. Only this part communicates with the Windows driver. A read request of the client is mapped to the corresponding interface functionality of the driver.

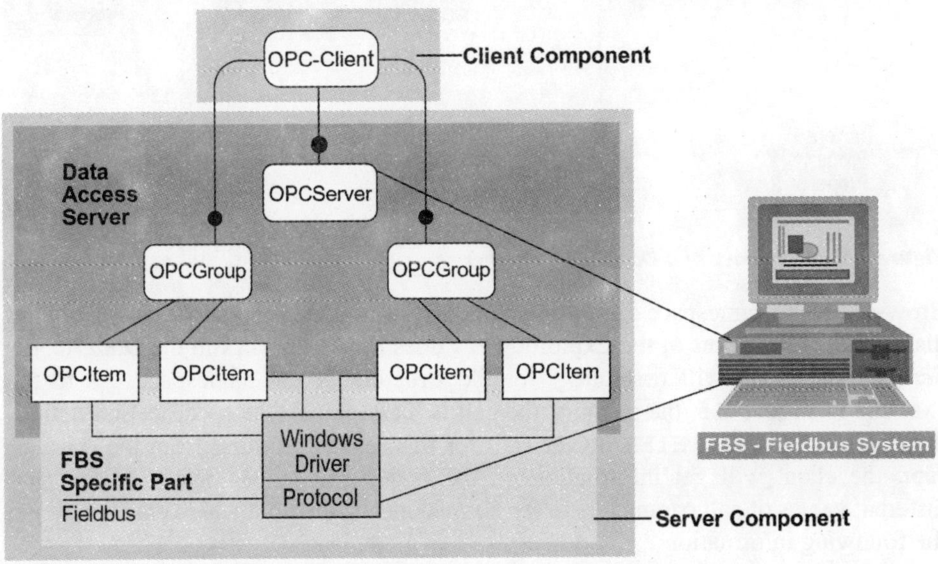

Figure 2.4: *Structure of a Data Access Server*

In the next part of this section, the functionality defined in the specification is described using specific scenarios.

Browsing of the namespace

After starting the server, the *Data Access Client* first accesses the top-level object OPCServer. The client browses the namespace to get information on its structure and to get identifiers which have to be used when creating OPCItem objects.

Figure 2.5 shows a part of a hierarchical namespace. The *Data Access Server* enables access to different measured values within a building. The namespace is created during server configuration. Other steps during configuration include the specification of the necessary communication parameters. The descriptions and identifiers of the items, properties, and AccessPaths can be freely chosen. As DCOM supports UNICODE (character coding which is supported by all languages), it is possible to create the identifiers of the namespace in any character set. A change of the implementation would therefore not be necessary if the *Data Access Server* was to be used with Japanese or Chinese character fonts.

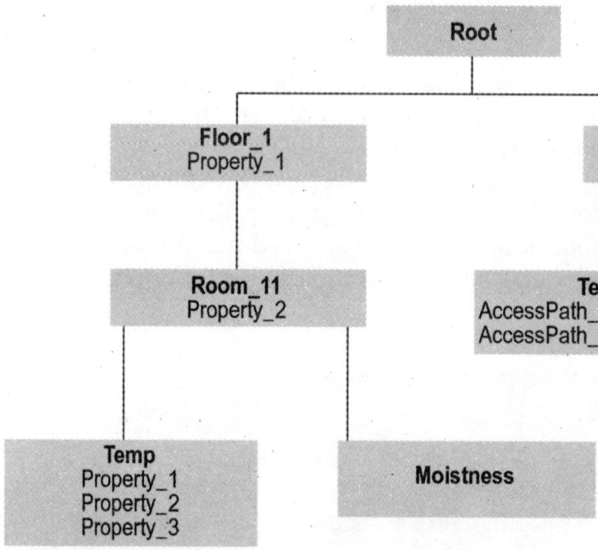

Figure 2.5: *Namespace of a Data Access Server*

Browsing of the namespace can be compared to a common process – browsing of hard disk contents by means of the Explorer. How does browsing work in the *Data Access Server*? The client will first query for the structure of the namespace. As to the example in **Figure 2.5**, the result of the call is hierarchical (the specification defines the constant `OPC_NS_HIERARCHIAL`[6]) for this result). Starting from the (virtual) root, the client will get information on the nodes and leaves below. It can pass different values of call parameters to the corresponding method, which enable it to get the following information:

- only nodes below the current node, (e.g. Room_11)
- only leaves below the current node, (Temp, Moistness) or
- everything below the current node As the return value for this call, the client receives strings containing all identifiers of the nodes and leaves below the current node. In this way, a flat namespace can be modelled in the client.

With regard to the leaves, a further restriction can be made with the following filters:

- only leaves the values of which have specific VARIANT data types (e.g. unsigned Integer 2-Byte)
- only leaves with specific access rights (e.g. only readable) or

6 This is not a spelling error in this book but in the specification.

– only leaves containing the corresponding string (e.g. Te*)

In a hierarchical namespace, the client has to be able to move through the namespace in order to get further information relating to a new node. A method supporting this navigation is available. The client is thus able to move from node to node. As the server now "knows" from which point the client calls other functions, it can return the corresponding values. If the client has moved to the node "Room_11" and requests the identifiers of the leaves below, it gets "Temp" and "Moistness" as a result. Is the client already able to use these results to create OPCItem objects? A look at the namespace shows that there is another leaf with the identifier "Temp." The client has to expand the identifier in order to make it unique for the server. This is achieved when the identifier of the node "Room_11" is added. The client itself does not have to do this; however, it can use the functionality of the *Data Access Server*. The latter returns *fully qualified ItemIds* (e.g. Room_11.Temp, where the delimiter "." is server-specific) which uniquely identify the items in the namespace. "Floor_1.Room_11. Temp" would be an example of such an identifier. The delimiters are always server-specific.

Creation of additional objects

After the *Data Access Client* has browsed the namespace, it will create OPCGroup and OPCItem objects[7]. OPCGroup objects are used for structuring OPCItem objects. This can be done for dynamic aspects (e.g. all quickly changing variables) or logical aspects (e.g. all variables in one specific area) or as desired. Nothing to this end is determined in the specification.

When requesting the creation of an OPCGroup object, the *Data Access Client* transmits values for the following parameters to the server:

– symbolic name

– *RequestedUpdateRate*

– *PercentDeadband*

– *ActiveState*

The symbolic name for the object can describe its semantic significance, e.g. "Temperatures_on_Floor_1." If the client does not transmit any value, the server will generate a unique name. The values of the parameters *RequestedUpdateRate*, *Percent-Deadband,* and *ActiveState* control the automatic data transfer to the client. The value for the *RequestedUpdateRate* influences the speed with which the values of the active OPCItem objects are scanned. With an inactive state of the OPCGroup object (*ActiveState* = FALSE), the process variables are not read automatically. The value of the *PercentDeadband* influences whether values are automatically sent to the client. A more detailed explanation of this functionality is made further below in this section. With each call of the method, the client is only able to create one single OPCGroup object.

7 This order is not determined by the specification. Of course, it is possible to create OPC-Group Objects first, then the namespace is browsed, and finally the OPCItem Objects are added.

In contrast, by calling the method for the creation of the OPCItem objects, the client can generate any amount of objects. By doing this, the *Data Access Client* passes values to the following parameters:

– *fully qualified ItemId* (unique path from the namespace)

– *ActiveState*

– *RequestedDatatype*

– *AccessPath* (optional)

– *ClientHandle*

The server returns values for the following parameter to the client after finishing the method call:

– *CanonicalDataType*

– *ServerHandle*

The *fully qualified ItemId* clearly identifies a data item in the namespace. This means that the *Data Access Server* can create an OPCItem object for this data item. Behavior of the object is the same as implemented for this data source/data sink. If the *fully qualified ItemId* describes the process variable "Temp" in the node "Room_11" from **Figure 2.5**, then the OPCItem object created for this purpose will support the reading of values from the corresponding device.

The state of an OPCItem object can be either active or inactive, which means that the object can be taken into account or not when automatically reading the process values.

The values in the namespace can have any desired VARIANT data type. A client may not support all these data types; it is possible that it does not implement any unsigned data types. Here, the concept concealed behind the parameter *RequestedDatatype* is used. When passing values for this parameter, the client can request a data type in which it wants to read or write the value. The server will then carry out a conversion during further calls. If the *Data Access Server* is not able to carry out the conversion required, it will inform the client by using a return parameter. Here, only the conversion from one VARIANT data type to another is concerned. Conversion of application data types into VARIANT data types (e.g. bits in Boolean) is server-specific and is carried out in the implementation of the OPCItem object.

What is the significance of the parameters *ServerHandle* and *ClientHandle*? What distinguishes them from the *fully qualified ItemId*? As already explained, the *fully qualified ItemId* uniquely identifies a leaf (item, variable) in the namespace. With the same *fully qualified ItemId*, a client can also create several OPCItem objects in different (or identical) OPCGroup objects. Several clients can create several OPCItem objects in one server component for the same *fully qualified ItemId*. In these two cases, the *fully qualified ItemId* is no longer sufficient for unique referencing of the OPCItem object, e.g. with read and write requests. For this purpose, *ClientHandles* and *ServerHandles* are used. As a pair, they are non-ambiguous for all OPCItem objects. The client will transmit different *ClientHandles* to the server with the request to create the OPCItem objects. Two clients can use the same *Client-*

Handle. The server allocates different *ServerHandles* in order to guarantee uniqueness on its part. Another reason for the existence of the handles is simplicity in accessing information on the OPCItem objects which client and server have to manage. The handle can be used as direct reference onto the information, as it is defined as a 32-bit value and has thus the same width as a pointer. Furthermore, programming can be effected more easily and more efficiently if a handle with the same size is used for addressing, instead of a string which has a different length for the different OPCItem objects. The client uses the server handle for deleting and changing object properties in the server. If the server transfers values to the client, the latter uses the handle for allocation.

The OPC client is able to delete OPCItem objects by invoking *RemoveItems*. Addressing is made via *ClientHandles and ServerHandles*.

The client can make the server check whether creation of OPCItem objects is possible with the parameter values provided. This protects it from unpleasant surprises, e.g. if a *Data Access Server* does not support a VARIANT data type conversion. The client can then react to this checking and request another conversion.

Access to Properties

Besides the real process variables, there is plenty of information in real applications that is important and interesting, such as the name of a device which is represented in the namespace by a node, or calibration possibilities for measuring devices, and a lot more. Mapping of this almost static information onto leaves in the namespace would unnecessarily expand it and would make browsing very time-consuming. For this reason, *Properties* have been introduced with version 2.0 of the *Data Access Specification*. They exist at nodes and leaves in the namespace.

The client is able to gain information linked to properties very easily. First of all, it queries for information to the properties available at one point (node or leaf) in the namespace. For every property, it gets a triple back. The triple consists of a numerical identifier (PropertyID), the data type of the *Property*, and the description in the English language.

The range of values for the *PropertyID* is subdivided into three subranges:

- 1–99 *OPC-specific Properties*. Definition of *properties* within this subrange is effected in OPC specifications. These properties have to be available at all items in the namespace. Several properties have been defined in detail up to now. Examples are *data type* and *access rights*.
- 100–4999 *Recommended Properties*. Definitions of properties within this subrange are provided in OPC specifications. These properties can be supported by a server. For example, with property 101 "Item Description" it is possible to deposit a more detailed description of the process variable, e.g. "Temperature at window in Room 11." Version 1.10 of the OPC Common Definitions and Interfaces specification defines ID subranges for different specifications.

- 5000–65535 *Vendor-specific Properties*. Here, it is possible to add properties for the manufacturers of the server, users, or others. For example, one can define properties that contain information on calibration possibilities.

In the next step, the client can query the value of the *Properties* by using their *PropertyIDs*.

The principal motive for using *Properties* is to make available static information. Should it be necessary, however, to write properties, the client can also determine a *fully qualified ItemId* for a property. By using these identifiers, OPCItem objects can also be created for properties.

Figure 2.6 represents the connection between the namespace and the OPC objects of a Data Access Server.

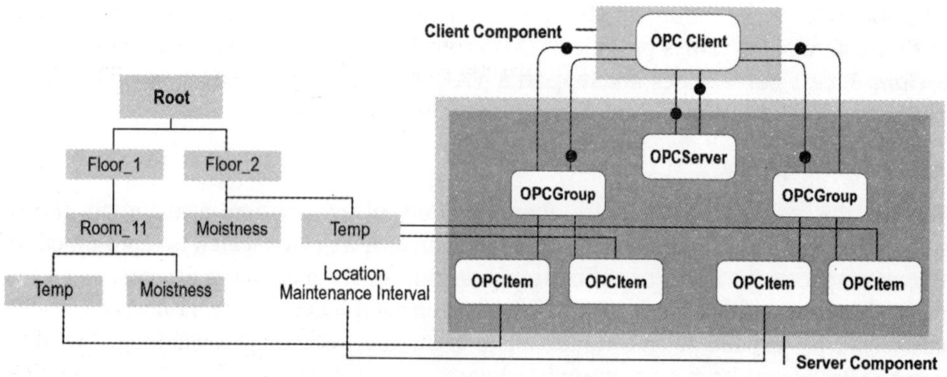

Figure 2.6: Namespace and Object hierarchy

On the left, you can see part of the namespace and on the right the OPC objects of the *Data Access Server*. The namespace is hierarchical. Besides the (virtual) root, it contains three nodes and four leaves which represent the process variables. One leaf has two properties, "Location" and "Maintenance interval." Two OPCGroup objects have been created in the *Data Access Server* which dispose of two OPCItem objects each. The OPCItem objects have been allocated items of the namespace by means of a *fully qualified ItemId* and thus process variables. A corresponding *fully qualified ItemId* could be: "Floor_1.Room_11.Temp". One OPCGroup object manages the OPCItem objects which represent temperature values. The other OPCGroup object contains a temperature value and information on the maintenance interval. The *OPC client* has access to the OPC objects in the server by methods at interfaces. This is illustrated in the picture by lines with black circles. Callback interfaces are also shown. For these interfaces, the black circles are closer to the client.

Reading and writing of values

After creation of the object hierarchy, the Data Access Client can read and write values or get information about the change of values.

Synchronous reading calls can be referred both to the cache and to the device. Cache reads are only possible if both the OPCGroup object and the OPCItem object are active since only in this case the cache is updated. The value of the parameter *RequestedUpdateRate* influences the freshness of the value in the cache as it controls the speed with which the server will read the value and enter it into the cache. With a Device read, the value is read directly from the device, i.e. from the data source.

Synchronous writing jobs can also be carried out. They are always directly effected to the device.

A further possibility of data exchange is available with asynchronous reading and writing requests. Here, the *Data Access Client* initiates the execution of the functionality. The method called immediately returns. The actual request is entered into a "logical" queue at the *Data Access Server*. After processing the request by the server, the client receives the result via callback, i.e. the server calls a method at an interface of the client. Association between request and response is made via a transaction identifier. The transaction identifier is a return parameter of the method called by the client. It is reused by the server when invoking the callback method.

With every synchronous or asynchronous call, the client can request values from one or several OPCItem objects or write to them. Addressing of objects is performed via Handles. With a large number of OPCItem objects in one OPCGroup object, many Handles may be transferred. In an extreme case, all handles may have to be transferred if all the OPCItem objects are to be read. For this case, a more efficient procedure is available.

With a *Refresh,* all active OPCItem objects of an active OPCGroup object are read. As data source, Cache or Device can again be selected. However, no explicit addressing of the OPCItem objects with handles is necessary.

Note that refresh applies only to reading data; and no "Write-Refresh" is available.

The method explained below is the most frequently used form of data exchange. Here, the roles of client and server are reversed. The client creates OPCGroup and OPCItem objects with corresponding parameter values (*PercentDeadband, UpdateRate, ActiveState*) which enable an automatic data transfer. Then, the server automatically reads the values of the active OPCItem objects for the corresponding active OPC-Group object, evaluates a condition, and sends the values to the client.

During definition of the namespace, the EU type (Engineering Unit) and EU information can be configured for individual process variables. If the process variable has a simple data type, the EU type "Analogous" and corresponding EU information, the following process will be carried out in the server:

The server calculates the absolute value of the difference of the EU information (30 in the example from **Figure 2.7**) once; it multiplies the result by the value of *Percent-Deadband* for an OPCGroup object (in our example 0.1, which results in 3). The server reads the active OPCItem objects as determined by the value of the *UpdateRate* (e.g. every 1000 ms). Furthermore, the absolute value of the difference between the value last sent and the currently scanned value of an OPCItem object is calculated in cycles. If the absolute value of this difference is greater than the value calculated only once (EU information, PercentDeadband), the currently scanned value of the OPCItem object is sent to the client. This is the case twice in **Figure 2.7** (values 48

...nd 52). The server carries out this calculation for all active OPCItem objects of the active OPCGroup object. All the values which meet the above-mentioned requirement are passed to the client, with one method call at the callback interface. The client identifies the individual values using both the ClientHandle and the ServerHandle. It is important to observe that the values *UpdateRate* and *PercentDeadband* are considered for all OPCItem objects of the OPCGroup object. Monitoring precision of the individual process variables has to be adjusted by means of the values of the EU information.[8] The effect of a smaller difference of the EU information, with the same value for *PercentDeadband,* is a higher sensitivity with regard to the changes of the value. Another possibility is to allocate this OPCItem object to another OPCGroup object with a different *PercentDeadband.*

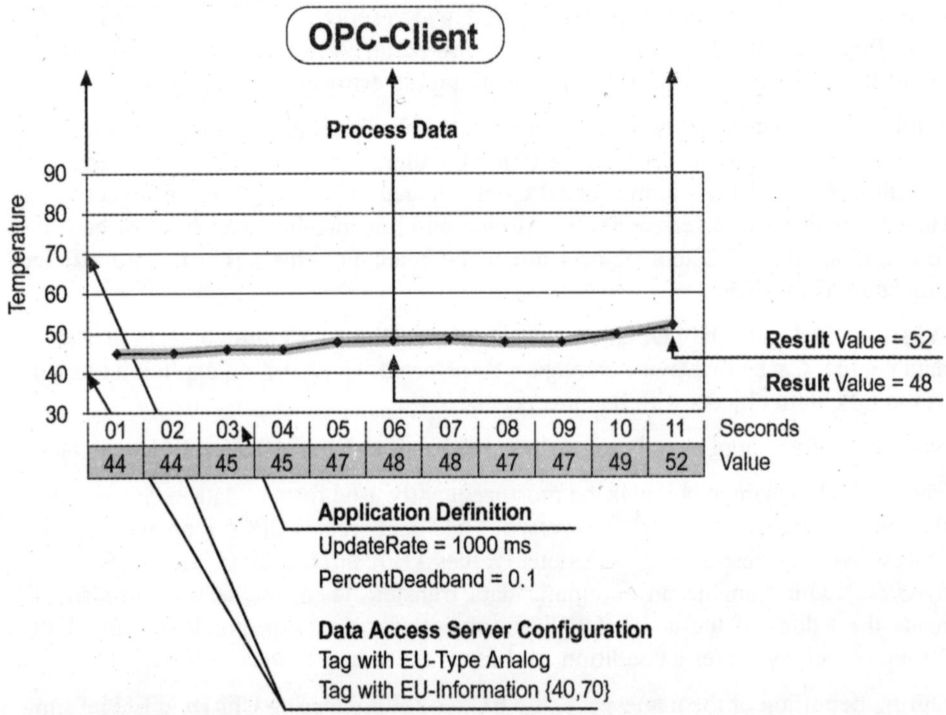

Figure 2.7: *Subscription-based data transfer – parameters and time flow*

This kind of data acquisition can only be used in the form described, with analogous values (EU Type "Analogous") of simple data types. It is not applied in the way described above in the following cases:

- With simple data types whose EU types are not equal to "Analogous" (the specification also defines "Enumerated"), the values are passed to the client every time they have changed.

8 This has been expanded by the OPC Data Access Specification 3.0, which allows defining *PercentDeadband* and *UpdateRate* at the OPCItem level.

– With values of structured data types (e. g. byte string), as the structural component to which the calculation refers cannot be stated. Structured values are passed by the server to the client every time they are changed.

Data are also automatically transferred to the client if the status related to the OPCItem object changes, e.g. from “Good” to “Bad.”

This completes the explanation of the functionality available for reading and writing. Most applications will only use this functionality. In the following sections, concepts are presented that cover special cases or allow for influencing the behavior of *OPC* objects after their creation (e.g. modification of the *PercentDeadband*).

Management of the life cycle of OPCGroup objects

Sometimes it is necessary or desired that OPCGroup and OPCItem objects exist without the client having an interface pointer to them. They are, for example, created by the client “as a reserve.” This is not possible under the specifications of DCOM, which relate the existence of an object to the access of its interfaces. This contradiction has been avoided by using a trick – when creating an OPCGroup object, its reference counter is set to the value 2. So it will have the value 1 if the client no longer has any interface pointers. This means that the object does not destroy itself if no interface pointers are used but has to be destroyed explicitly. To do this, the method *RemoveGroup* is used.

What is the advantage of this concept and how can it be used?

A client can create a set of OPCGroup objects with the corresponding OPCItem objects and then release all the interfaces. The objects will remain. If the client needs to access the objects at a later time, it has to first obtain an interface pointer to the OPCGroup object and information on the OPCItem objects. This is done in several steps. The client will first of all create an OPCGroup-*Enumerator*[9]. This contains information on the *OPCGroup* objects in the *Data Access Server* component. By using the name of the OPCGroup object, the client can request a method that returns an interface pointer to the object. At the OPCGroup object, a method is again available that generates an enumerator for the OPCItem objects. Now the client can obtain all the necessary information on these OPC objects (e.g. ClientHandle and ServerHandle).

Public and Private Groups

Another concept is to distinguish between public and private groups. After a client has created an OPCGroup object, it belongs to it; it is private. It represents a certain view onto the process. Sometimes it is useful and necessary to make the same view available to other clients. To do this, the client changes its private group into a public

9 Enumerator objects are specific DCOM objects. At the time they are created, they store the current status of the objects in the server, e.g. all property values of the OPCGroup objects. Modifications of the status at a later date, e.g. modification of the UpdateRate of an OPCGroup object, are not taken over by the enumerator objects. Enumerator objects have an interface with methods that are modified for reading out the status through the client. At some places in the Data Access Specification enumerators are specified.

group, which is visible and which can also be used by other clients. A further possibility for the creation of public groups is that the server makes them available immediately after launching.

A client interested in access to public groups will use the following procedure:

It creates an OPCGroup *enumerator*, requests the name of the public group(s), calls a method which returns the interface pointer to the desired OPCGroup object, and then queries for the properties of the OPCItem objects managed by the OPCGroup object. The client can adapt "its" public OPCGroup object to special requirements by changing the values of *UpdateRate*, *PercentDeadband,* and *State* for instance. OPCItem objects can neither be added to public groups nor deleted. The client, however, is able to change their *State* and the *RequestedDataType*.

If a client wants to carry out structural modifications at a public group nevertheless, it has to call the method *CloneGroup* at an interface of the OPCGroup object to create a private copy. Modifications to this private OPCGroup object are not, however, reflected in the initial form of the public Group object and in instances derived from it.

Status information and influencing of different objects

There are a large number of methods with which the client can request and influence the status (behavior) of OPC objects. A short overview follows.

OPCServer object

When requesting the method *GetStatus* the client obtains information on the parameters shown in **Table 2.3**.

Table 2.3: *Return values of the GetStatus request at the OPCServer object*

StartTime	This value contains the time when the server was started
CurrentTime	This value contains the time in the server when the method call was made
LastUpdateTime	This value indicates when the last data were sent to the client by callback
ServerState	This parameter indicates if the server is in the status running, failed, noconfig, suspended or test
GroupCount	This parameter indicates the total number of OPCGroup objects which the clients have created in the server
BandWidth	This parameter indicates the processor performance consumed by the server
MajorVersion, MinorVersion, BuildNumber	These parameters indicate the version status of the server
VendorInfo	This parameter gives information on the manufacturer

The status of the OPCServer object cannot be influenced directly. It is thus not possible to set the time.

OPCGroup object

The client can query the properties of an OPCGroup object by means of *GetState*. It then gets information on the parameters described in **Table 2.4**.

Table 2.4: *Return values of the GetState request at the OPCGroup object*

UpdateRate	This parameter indicates the number of times the values for the active OPCItem object have to be read
ActiveState	The value can be TRUE or FALSE. Only with TRUE, values of the OPCItem objects are read
Name	This parameter indicates the symbolic name of the object
TimeBias	This parameter indicates the deviation from GMT (Greenwich Mean Time) in minutes
PercentDeadband	This parameter defines when values are automatically sent to the client (see fig. 2.7 and description).
LCID	This parameter indicates the language identifier valid for this object
ClientGroup- and ServerGroupHandle	These parameter values are used for the client or server internal addressing

By using *SetState*, the client can change these properties. The symbolic name of the OPCGroup object can be changed by means of *SetName*.

OPCItem object

After creating the OPCItem objects, the client can influence the behavior of the OPCItem objects with the following methods:

By using *SetClientHandles*, the client can change the *ClientHandle*. If the client wants to influence the *RequestedDatatype*, it will use *SetDatatype*. The client is able to influence the state by calling *SetActiveState*.

Different possibilities of implementing the namespace

For creating the OPCItem objects, the client has to pass the *fully qualified ItemId* to the server. This identifier provides a unique reference to a variable and can be obtained as the result of browsing the namespace. This functionality and, thus, the existence of a namespace is optional, however. If no namespace exists, the client must be provided with another possibility to get the complete *ItemID*. A convenient way to do this is to define a syntax and a semantics for the server. The user passes a corresponding string to the client, which passes the *ItemID* to the server. The server then checks the validity and usability of the *ItemID*.

When manually entering the *ItemID*, the user has to know the relevant syntax and the semantics and has to avoid input errors. If the *ItemID* is determined by browsing the namespace, the namespace has to be configured. Users may consider this unnecessary if a comparable structure, e.g. number of variables of a PLC program, already exists. Which of the two implementation possibilities is preferred will depend on the specific application area. Server manufacturers should take advantage of the flexibility OPC offers to provide optimum ease of use.

Loading and storing of configuration information

Is it possible for the *Data Access Server* to load or store configuration information about the process data? This would for instance include communication parameters. Yes, this possibility exists. At an optional interface of the OPCServer object, functionality is available with the help of which a file can be loaded and stored. The information contained in this file, however, only describes how to access the process variables and does not contain any information. This concept, however, has some deficiencies:

- The client gets no information on whether configuration of the server based on the loaded file is finished, i.e. whether it is possible to browse the namespace.
- The client gets no information on any errors that occurred during configuration.
- It is not specified what happens when different clients load various configurations at the same time.

Some of these can be avoided in the following way:

A component of the server status allows the server to show that it needs configuration information (OPC_STATUS_NOCONFIG). When configuration is finished, the server is able to show this with the status OPC_STATUS_RUNNING. To get this information, the client has to constantly query the server status. For individual items in the namespace, the quality "BAD Configuration Error" can later indicate, when reading or writing, that configuration for this data item has not been successful. Information, however, should be given much earlier.

Figure 2.8 shows the OPC objects with their interfaces which are defined in the *Data Access Specification*. During this definition process, some details have to be observed:

- The symbolic name of an interface always starts with the capital letter **I**.
- If interfaces are concerned which were not defined by Microsoft, the first letter should be followed by an abbreviation describing the field of application. In the case of the OPC specifications, this is **OPC**.

A further requirement is to always enclose optional interfaces by square brackets. These labelings are, of course, only of a symbolic kind. Reference at runtime is made via the numeric interface identifier.

The object OPCServer is the Top-Level-Object in the object hierarchy of the *Data Access Servers*. It is created when the *Data Access Server* is launched and a pointer to a desired interface is passed to the client. An OPC client can use the functionality available at the interfaces. **Figure 2.8** shows all the interfaces at the OPCServer object. Besides the OPC interfaces, there are also interfaces specified by Microsoft. Furthermore, there are three optional interfaces.

Figure 2.8 also illustrates the OPCGroup object with its interfaces. At this object there are interfaces defined by Microsoft, OPC interfaces, and optional interfaces.

We have already mentioned several times the existence of Callback Interfaces at the client. There are two Callback Interfaces:

- *IOPCDataCallback* with methods for notification on the completion of asynchronous method calls and for automatic passing of data to the client.
- *IOPCServerShutdown* for information on an imminent shutdown of the server. An explanation of this functionality has already been given in section 2.2.2.

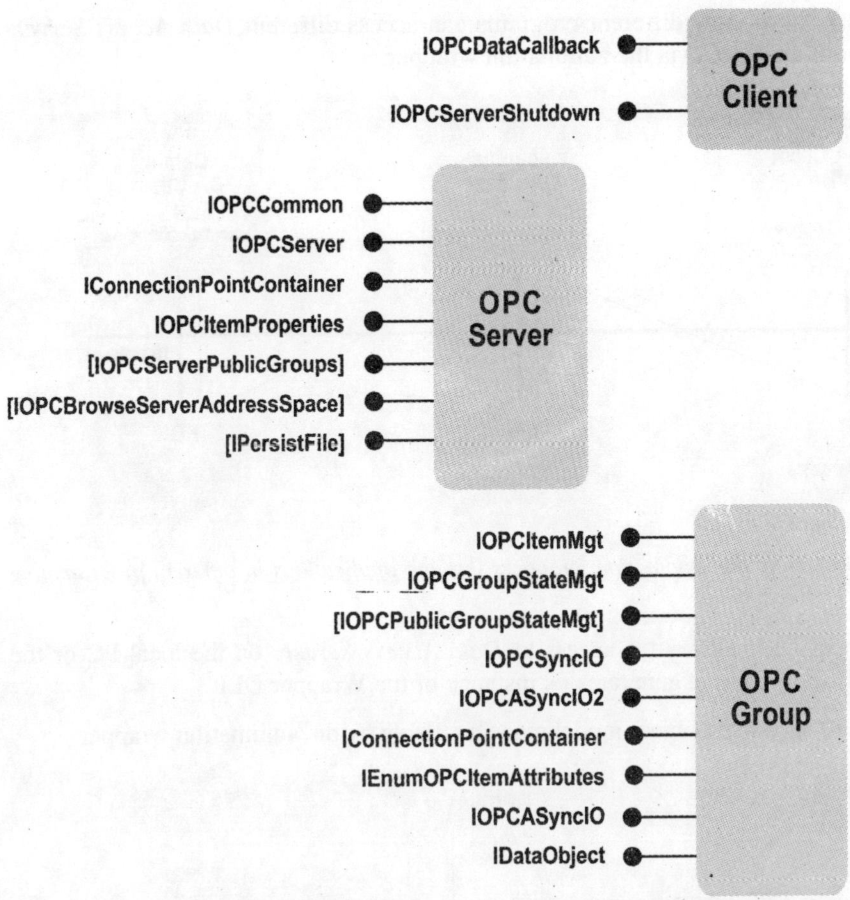

Figure 2.8: OPC objects in the Data Access Server and Client

Automation Interface

Section 3.2 describes how a client accesses the functionality of a server. To do this, it needs a reference pointer to an interface. This access type is particularly useful for languages that support pointers, such as C/C++. Other languages like VisualBasic do not support pointers. VisualBasic was designed for easy implementation of applications. VisualBasic users are to be relieved of having to deal with the many details of DCOM programming. These tasks are performed by an automation controller, which is part of the runtime environment of the language. A server has to support an automation interface for this purpose. This automation interface additionally gives access to auxiliary functions that need to be implemented by the server. The OPC Foundation provides the automation wrapper for this purpose. It implements the corresponding interface and maps calls to the custom interface. In this way, server manufacturers do not all have to implement the same functionality. The goal for a long time had been to make automation wrappers available for all specifications. This goal has been dropped meanwhile. An automation wrapper is today only provided for the Data Access Specification 2.05.

Figure 2.9 shows how different programs can access different *Data Access Servers* with custom interfaces via the automation wrapper.

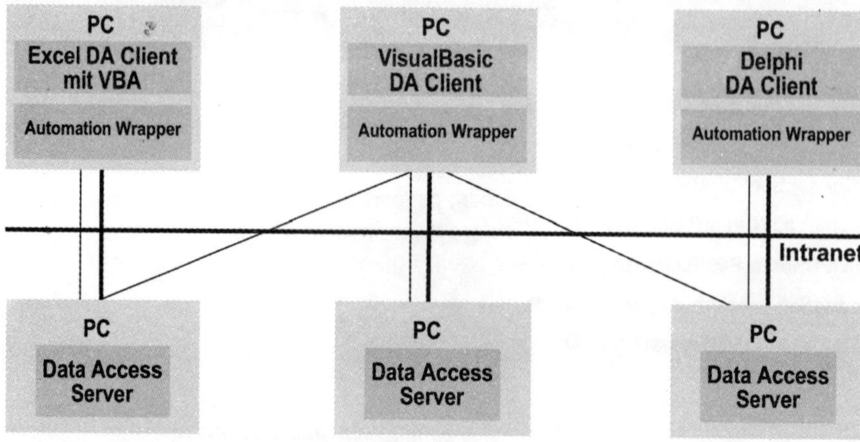

Figure 2.9: Use of the automation wrappers through applications in different programming languages

The wrapper DLL allows for access to *Data Access Servers* on the local PC or the remote PC. Every client generates its instance of the Wrapper DLL.

Figure 2.10 shows the object model made available by the automation wrapper.

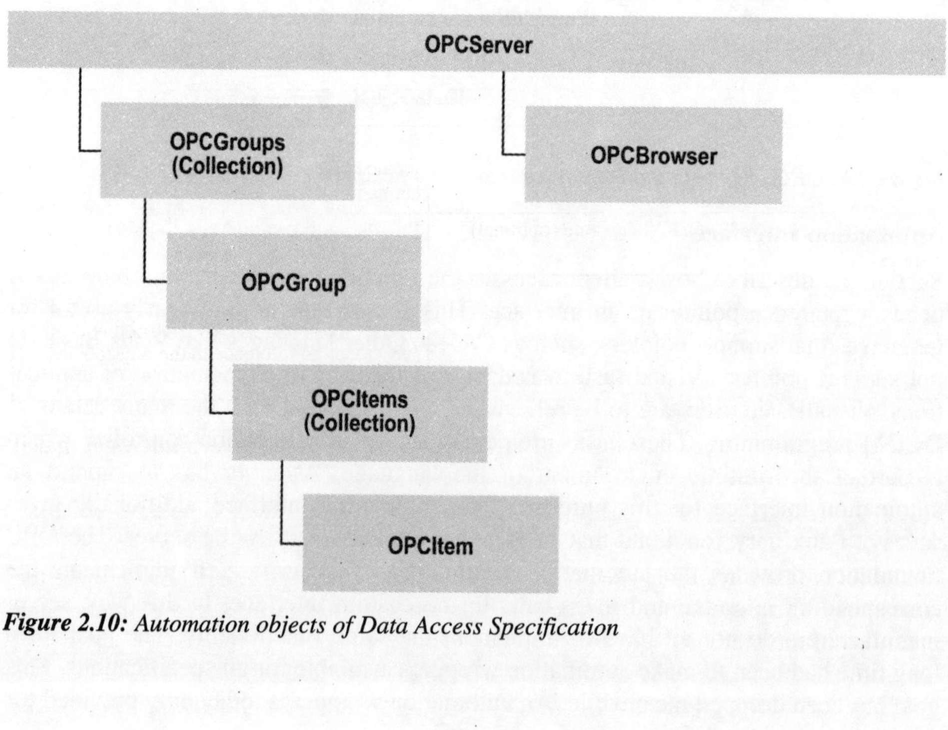

Figure 2.10: Automation objects of Data Access Specification

There are three clear differences between this model and the object model for the custom interfaces of **Figure 2.2**:

- There is a single automation object which makes available the browser functionality for the client. If a server does not support browsing, the object does not exist.
- There are collection objects which exist between the objects at intermediate stage.
- There is an automation object OPCItem, i.e. a client has direct access to the methods at this object.

All the information on the objects, methods, properties, and events of the automation objects of the *Data Access Specification* are contained in a type library and can thus be easily used during development. This is why these definitions are not explained in detail in this section.

OPCServer object

The object makes available a variety of properties which correspond to the method calls *GetStatus*, *Get/SetLocalID,* and *SetClientName* at the custom interface. Further properties are also available.

The object makes available all the methods existing at the custom interfaces of the OPCServer object, with the exception of the methods for browsing the namespace and for adding or removing of OPCGroup objects. They are made available either at the OPCBrowser object or at the OPCGroups object.

In addition, the methods shown in **Table 2.5** are available.

***Table 2.5:** Methods at the OPCServer automation object*

Method	Parameter	Meaning
GetOPCServers	NodeName (optional)	Return the ProgID of existing servers. This method accesses functionality of the server enumerator (see chapter 2.2.2).
Connect	ProgId, NodeName (optional)	Provides a link to the Data Access Server which implements Custom Interfaces
Disconnect		Releases the link to the server
CreateBrowser		Creates a browser object. If browsing is not supported, no reference to the object is returned.

The *Data Access Server* can inform the client via the Event *ServerShutDown* that it will soon shut down.

OPCBrowser object

This object makes available methods and properties allowing the client to browse the namespace. There are some differences to the custom interface as to the number and semantics of the methods.

OPCGroups object (Collection)

There is no comparable object with custom interfaces available. The collection object is used for adding, managing, and removing OPCGroup objects. It also has properties. Methods for adding OPCGroup objects are available to the client. There is an event at the object which informs the client of the occurrence of a value or state change at any OPCGroup object.

OPCGroup object

The OPCGroup object provides access to (VisualBasic) properties. This object offers all the methods also available at the Custom interfaces of the OPCGroup object. There are differences in the methods for the synchronous and asynchronous data exchange. There are no interfaces at an automation object. It is possible to use methods with the same name at different custom interfaces as distinctions are made via the interface pointers. Different names are used at the automation object:

- SyncRead and SyncWrite for the synchronous data exchange.
- AsyncRead, AsyncWrite, AsyncRefresh, and AsyncCancel for asynchronous data exchange.

There are events which inform the client of the completion of asynchronous calls or about value changes.

OPCItems object (Collection)

This object is a collection object for the OPCItem objects. The collection object is used for adding, managing, and removing OPCItem objects. A large number of properties are available. The properties *DefaultRequestedDataType*, *DefaultAccessPath*, and *DefaultIsActive* contain default settings that are used when adding OPCItem objects.

OPCItem object

Objects of this kind form the lowest level in the hierarchy of automation objects. The object has properties. The client is able to read and write the value of the object via the methods Read and Write. No events are available at this object.

2.4 OPC XML-DA Specification

2.4.1 Field of application and model

Work on the OPC XML-DA specification started in the year 2000 with the aim of integrating Microsoft's XML activities into OPC. This objective was later extended to making the specification applicable independently of the computer platform and location. OPC XML-DA products can be used on both the intranet and the Internet. They allow implementation into different environments supporting HTTP and XML.

A look at the framework conditions soon showed that the new specification could not be fully modeled on the OPC DCOM DA specification for various reasons:

- The OPC DCOM DA specification defines an object hierarchy (OPCServer, OPC-Group, OPCItem). In a heterogeneous environment, however, computers cannot all be expected to implement the corresponding object models. This also applies to launching and shutting down the servers as well as to scanning for them. OPC DCOM either provides DCOM-specific solutions for these tasks or uses DCOM directly. For OPC XML other, generally applicable solutions have to be found.
- Data transmission with XML is less efficient than the transmission of binary data with DCOM. Therefore, the client should have the possibility to choose between the amount of information and the transmission performance. A large number of methods are available at the OPC DCOM DA interfaces. To increase efficiency, the combination of multiple methods is useful.
- OPC DCOM DA defines callbacks. This definition, too, is linked to requirements that do not necessarily exist in the context of Web services. HTTP is a stateless protocol. The server does not have a lasting knowledge about the client. Some HTTP extensions allow storing the relevant data, but these HTTP extensions are not always supported. As explained earlier in this section, servers (and clients) cannot always be expected to implement object models. Consequently, the end points for the callbacks would be missing.
- In the DCOM-based OPC world, there is the server enumerator. The client can use it to find the existing servers. A similar infrastructure that allows scanning for OPC XML-DA servers is not available (yet).

2.4.2 Specification object and use

This section describes the methods you find listed in **Table 2.6.**

The GetStatus call can be used for checking whether the OPC XML-DA server is available. If the client receives a corresponding response, it knows that the server is available. The client can also use this call to query information about the server.

Table 2.6: *Methods of the OPC XML-DA specification*

GetStatus/GetStatusResponse	Query server status
Browse/BrowseResponse	Query namespace, including property information
GetProperties/GetPropertiesResponse	Query property values
Read/ReadResponse	Read data
Write/WriteResponse	Write data
Subscribe (Client)/SubscribeResponse (Server)	Define subscription (by client)
SubscriptionCancel/Response	Cancel subscription (by client)
SubscriptionPolledRefresh(Client)/ SubscriptionPolledRefreshResponse(Server)	Client-initiated polling for subscription values

The namespace of an OPC XML-DA server also contains all the variables that the client can read-access and/or write-access. The large number of methods defined in

OPC DCOM DA for browsing the namespace have been combined into a single method. There is one call for the client browse. The server responds with a BrowseResponse. The client can specify filters, the maximum number of items to be returned, and the browse start and continuation points for repeated calls.

With the GetProperties call, the client can query the values for the properties. The client receives the information on the existing properties in the BrowseResponse.

After the client has browsed the namespace, it can read and write values. It passes the complete ItemIdentifier directly for the required variable.

The behavior for writing the data is similar to that for reading. The client passes the complete ItemIdentifier and the value to be written. The server transfers the value and returns the result to the client. Besides the value, the client can also write the timestamp and the quality information as well as read back the value.

The specification also offers the possibility of subscription. Web services do not support callbacks. Therefore, a different way had to be found. The following explains the possibility shown in **Figure 2.11**:

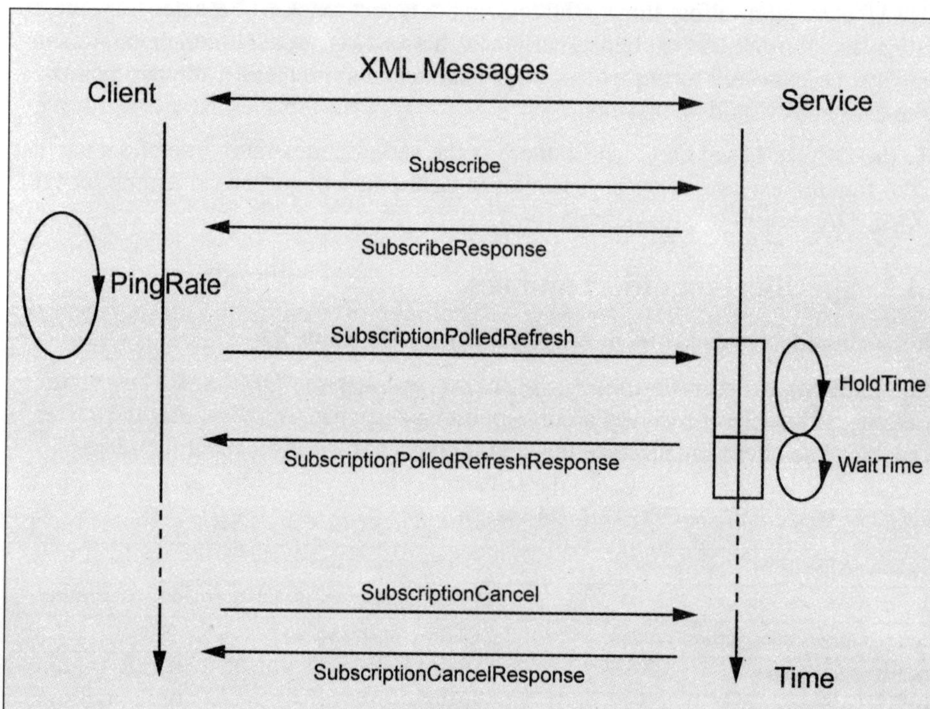

Figure 2.11: *Sequence of automatic data transmission based on the OPC XML-DA specification*

To open a subscription, the client sends a request. This request indicates the variables the client is interested in. Various values, e.g. for RequestedSamplingRate, Deadband and EnablingBuffering, can be preset. The RequestedSamplingRate corresponds to the UpdateRate of OPC DCOM DA. The EnableBuffering parameter has the following meaning: the client cannot query the values of the callback as often as the

RequestedSamplingRate allows. This means that the server gets the values more frequently than the client queries them. If the values that the server just got change and if EnableBuffering=TRUE, the server will store the values and transmit them to the client with the next call.

The SubscriptionPingRate is a timeout preset by the client. This timeout is monitored by the server. If the client's calls for fetching the values are not faster than the SubscriptionPingRate, the server will close the subscription. With the callback rate, the client informs the server about the frequency at which it will query the acquired values.

The server returns a SubscriptionResponse. It contains the callback handle, the supported callback rate, and the available item values.

To close a subscription, the client calls SubscriptionCancelRequest. The server returns a SubscriptionCancelResponse.

Picking up the data for a subscription is initiated by the client. It calls Subscription-PolledRefresh and receives the data with SubscriptionPolledRefreshResponse. In the request, the client can tell the server how long it will wait for the response. In this way, the client can indicate that the connection is to remain open for a specific period of time. By specifying the values for HoldTime (absolute point in time) and WaitTime (period of time), the behavior of an automatic transmission of the values from the server to the client can be reproduced. The server waits for at least the HoldTime before transmitting the response. If the valuues do not change within the HoldTime, the server continues waiting. Here is where the WaitTime comes in. In this case, however, the server immediateley transmits the response to the client as soon as a change occurs. The cycle is then complete. If the period set by WaitTime expires without a change in data occurring, an empty response is transmitted.

The specification defines the following possibilities of adapting the data exchange between the client and the server to different requirements:

- HoldTime (absolute point in time) and WaitTime (period of time), the behavior of an automatic transmission of the values from the server to the client can be reproduced. The server waits for at least the HoldTime before transmitting the response. If the values do not change within the HoldTime, the server continues waiting. Here is where the WaitTime comes in. In this case, however, the server immediately transmits the response to the client as soon as a change occurs. The cycle is then complete. If the period set by WaitTime expires without a change in data occurring, an empty response is transmitted.
- With MaxElementsReturned, the client can limit the amount of item information returned during browsing. For the next call, the client passes the ContinuationPoint received from the server.
- When the client calls Read, it can pass values for MaxAge to specify up to how old a value may be. If it passes MaxAge = 0, the server has to read the value from the source. If MaxAge>0, the server will compare if a cached value is recent enough. If it is, this value will be passed; if it is not, the server will read the value from the source.

– When writing values, the client can request that the value be read back. This is done in the form of the information accepted by the server.

– With OPÇ DCOM DA, the client uses the complete ItemId for requesting the creation of an OPCItem object. OPCItem objects are then addressed via handles. In OPC XML-DA, there are no OPCItem objects. Therefore, the variable of interest always has to be addressed by using the symbolic identifier. For efficiency reasons, this addressing is split up into ItemPath and ItemName. This means that the complete identifier is not transmitted for each item. For requests, settings can be defined at different levels. For example: MaxAge can be specified for the entire request, for a list of items or for a single item. This can also be done in one call. The concrete parameter overwrites the less concrete one, which means that ItemLevel takes precedence over ListLevel, which takes precedence over RequestLevel.

– The server can be informed in a call that particular information does not have to be transmitted in the response. Examples are time data, quality data, etc.

– The server can pass values for RcvTime and RplyTime to the client in the response. The client can evaluate the values and draw conclusions on transmission and processing times, for example.

2.5 OPC Data eXchange Specification

2.5.1 Field of application and model

What factors made it necessary to define the OPC DX specification?

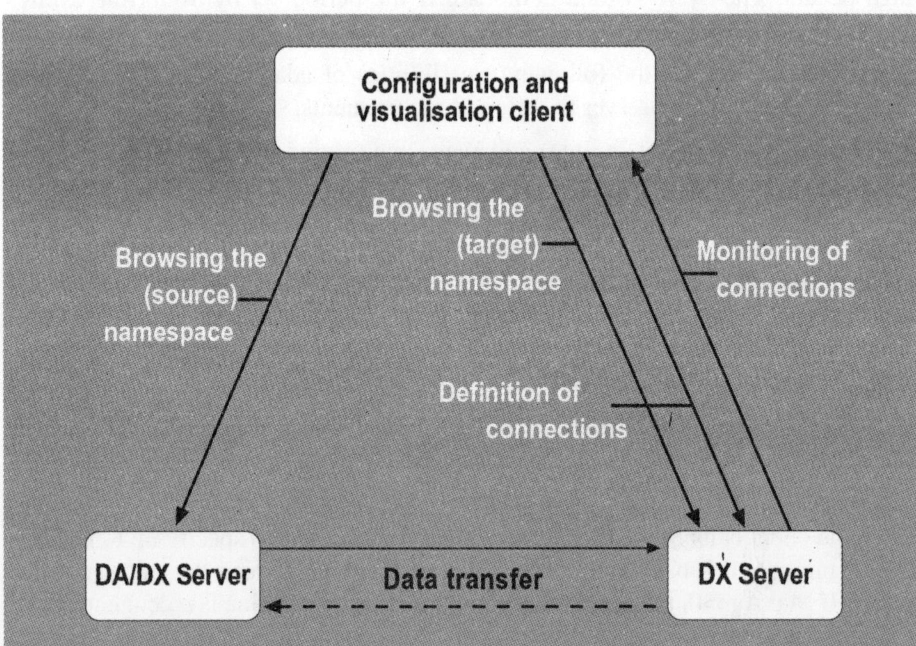

Figure 2.12: *Structure of an OPC DX application*

As mentioned earlier, OPC has a huge installed basis. OPC Data Access servers may sometimes need to exchange data directly, for example if information has to be transmitted from one production area to another. This currently requires the use of clients or proprietary bridges, which makes the data exchange slow and excludes generally applicable solutions.

In addition, a large number of fieldbus systems use TCP/IP on Ethernet as the transport and network protocol. Even though they all use the same medium, they cannot communicate directly because they use different application protocols. OPC DX aims at providing a solution to this problem, as well.

Figure 2.12 shows the structure of an OPC DX application. A DX server can receive data from one (or more) DA server(s) or one (or more) DX server(s). DA client functionality is implemented into the DX server for this purpose. With this client, the DX server creates OPCGroup and OPCItem objects. The data point from which the data are acquired is called the source item, and the data point to which they are written is called the target item. This link and its property is called a connection in the specification. A DX configuration consists of the total of all connections. Connections are defined by a configuration client and are mapped to the namespace of the DX server. A monitoring client monitors the connections. This is done with the help of Data Access functionality. Other clients (visualization) can access the existing items "as usual".

2.5.2 Specification object and use

The specification distinguishes the configuration model and the runtime model. The following section will explain the different aspects of these two models.

Configuration model

The configuration model defines the connection semantics as well as the ways in which the connections are created, deleted, modified and monitored. The information about the connections is stored in a defined area of the namespace. The attributes of a connection – and thus the structure of this namespace area – are specified. A Data Access client can browse the namespace and have OPCItem objects created for variables it is interested in. Besides the DA client functionality, a DX server also implements DA server functionality.

A configuration client can use different commands to create, modify and delete connections.

Runtime model

Besides the configuration, the specification also defines the behavior of a DX server at runtime, that is during horizontal data exchange. The runtime model deals with the data transfer between DX and DA/DX servers. It defines how to acquire data, monitor the status, and detect errors.

The DX server can subscribe to values from a Data Access server that supports one of the specification versions 2.04, 2.05 or 3.0. Another possibility is that the DX server supports the OPC XML-DA specification and receives data from OPC XML-DA servers.

The target item and the source item can be of different data types. When creating the OPCItem object, the DX server can request the DA/DX server to do this conversion. If this is not supported, the DX server has to convert the data itself. Future extensions to the conversions (e.g. from Celsius to Fahrenheit) are being considered. They will then be defined as a property of the connection.

The specification also defines the behavior of a DX server on detection of an interruption of the source server connection. In this case, the server will try to reopen the connection at the frequency of an adjustable ping rate. If no connection to the source server exists, this information is mapped to the corresponding attributes of the target item. Depending on the properties of the connection, the DX server can preset a substitute value instead of the actual value.

A monitoring client can subscribe to the corresponding items in the configuration branch of the namespace to get information on the connection states.

The DX server can store the entire current configuration at runtime. This means that all results from the calls for adding, modifying and deleting connections are saved. How and where this is done is not defined in the specification. When the DX server is launched, the current configuration can be reloaded and is then available again.

2.6 OPC Complex Data Specification

2.6.1 Field of application and model

A *Data Access Server* supports access to simple and complex data. A temperature value is an example of simple data, whereas records of diagnostic information serve as examples of complex data. Both data types (simple and complex) can be read or written in OPC DCOM and XML *Data Access* applications. Yet, the OPC specifications described previously in this chapter do not define the means for client and server to exchange structural information about complex data.

Besides the means of getting the type information, the following aspects need to be considered with regard to the specification:

- The client understands the structure of the data. In this case the client may forward individual elements of the complex data item.
- The client understands the structure of the data and knows its semantics. In this case the client is not only able to distinguish between elements but also knows about their type and relations to each other.

2.6.2 Specification object and use

The specification proposes two approaches:

- type descriptions defined within OPC specifications

– type descriptions defined outside OPC specifications (e.g. Fieldbus organizations)

OPC type descriptions are concerned with the definition of the structure of complex data. This allows the client to know the elements of complex data items. The type description is effected using XML. This description system is not very flexible, but it provides for easy implementation.

The type description systems of other organizations can also contain semantic definitions and allow a client to know the semantics. These descriptions can be more flexible but they increase the requirements for implementation, as client and server should be able to understand a number of type description systems.

The specification defines that information on and about complex data items must be provided using defined properties. **Table 2.7** shows the defined properties and their meanings.

***Table 2.7:** Properties for the description of complex data*

PropertyId	Name	Meaning
109	Complex Data Type Description System	Identifies the type description system used (e.g. OPC, Fieldbus Consortium)
110	Complex Data Type Description ID	Identifies the type description of a complex data item (e.g. reference to a XML file, refe rence number related to a type description system of consortia)
111	Complex Data Type Description Version	Identifies the version of the type description
112	Complex Data Type Description	A BLOB that contains the information necessary for clients to interpret the value of the complex data item

Only items with complex data types must support these properties.

Besides the possibility of describing complex data types, the new specification also defines the behavior during writing of complex data. Not all elements of the complex data item may be writable. In this case, the client supplies the entire buffer in a write request and the server ignores the values of non-writable elements.

In a case where the value for one or more writable elements is invalid, or cannot be applied, the server rejects the entire write request.

2.7 OPC Command Execution Specification

2.7.1 Field of application and model

In many applications, it is not only the reading and writing of values that is important, but also the execution of commands. Basically, commands can also be mapped to items. A client then has to write specific values and evaluate the values read. This approach does not fully correspond to the semantics of commands. Commands may well be executed in parallel. Different input parameters can be used in this regard. The command execution is often based on state machines. State transitions are therefore of special importance. Commands are not only important in connection with Data Access. They can also play a significant part with regard to alarms or historical data. Considering these facts, it makes sense to define a separate specification for the execution of commands.

The OPC Command specification is based on a model that considers these requirements. There is functionality for finding the available commands and the corresponding state machines. Commands are structured in domains and assigned to targets. XML is used for describing the command properties. The client can have commands executed, and it can monitor the execution. The full functionality is available both as a COM interface and as a Web service. The OPC Commands specification is a self-contained specification, which can be used as a basis for developing standalone clients and servers. The definitions can also be implemented together with DA, AE, HDA, etc.

2.7.2 Specification object and use

The client uses command information services to get information on the available commands and properties of the command server. Commands are structured in domains and are connected to targets. Examples of domains are device configuration or program invocation. In that case, a specific automation device represents a target.

The client can request information on domains and the commands contained. It can also query the assignments between targets and commands. Descriptions are available for the existing commands.

Table 2.8: *Command information services*

QueryCababilities	Query of the OPC command server properties
QueryCMD	List of the domains and their commands
BrowseCMDTargets	List of the targets with the associated descriptions as well as the domains
GetCMDDescription	All the information about a defined command (states, actions, events, etc.)

When the client has obtained the information about the available commands, it can execute them. This can be done synchronously or asynchronously. With asynchronously executed commands, the client can request notification in the case of status changes.

Table 2.9: *Command execution services*

SyncInvoke	The command is executed synchronously. Status changes are transmitted as the return values
AsyncInvoke	The command is executed asynchronously
Connect	The client can connect to an asynchronously called command

2.8 OPC Data Access Specification 3.0

2.8.1 Field of application and model

The OPC Data Access Specification 3.0 is a further development of the 2.05 version. Most of the innovations resulted from work on the OPC XML-DA and OPC DX specifications. Some solutions of these specifications have proven better than the previous definitions. For example:

- Browsing the namespace was simplified considerably.
- Item values can be accessed directly without the previous creation of an object hierarchy.

2.8.2 Specification object and use

What innovations are there and what consequences do the changes have?

New interface IOPCBrowse

There is only one call for browsing the namespace. Navigating in the namespace is no longer necessary. An additional call is available for accessing property values. Therefore, IOPCItemProperties is not supported any longer.

New interfaces IOPCItemDeadbandMgt and IOPCItemSamplingMgt

Deadband and SamplingRate now can also be defined at item level. In an application, this may result in several values being transmitted for an item to the client with the callback, if the GroupUpdateRate is greater than the SamplingRate.

New interface IOPCItemIO

The possibility to write and read values without creating an object hierarchy can be useful for applications in which as little memory capacity as possible is to be used.

New interface IOPCGroupStateMgt2

Connection monitoring is now available for subscriptions. On the one hand, this meets a frequently demanded requirement. On the other hand, it has been influenced by the definitions of the OPC DX specification.

New interfaces IOPCSyncIO2 and IOPCAsyncIO3

Writing the quality and the timestamp corresponds to the functionality of the OPC XML-DA specification. It is now also possible to read values of a maximum age. This means that the server uses the value set in MaxAge to decide whether to get the values from the CACHE or from the DEVICE.

At the same time, the support of particular interfaces is no longer demanded in implementations. This applies to interfaces that had been optional anyway. Interestingly enough, the optional IOPCBrowseServerAddressSpace has been replaced by IOPCBrowse, which is not optional.

2.9 OPC Alarms and Events Specification

2.9.1 Field of application and model

The explanations in this section relate to version 1.10 of this specification.

The *Alarms and Events Specification* defines an interface between client and server programs for structuring, monitoring, transmitting, and acknowledging events and alarms. The *Alarms and Events Server* can record and evaluate values from different data sources and decide whether an event has occurred. The data sources can be located directly on I/O cards plugged in into the PC. However, they can also reside on any devices, such as regulators and input/output modules, which are connected via serial links or via field buses.

Alarms and Events Server and *Data Access Servers* may have the same data sources. What is the difference between the two?

Data Access Servers make available a relatively continuous data flow. The automatic transfer of values can be adapted to relative value changes (e.g. +/-1 unit) by means of a *PercentDeadband*. This possibility of adaptation only exists if these values are analogous variables (EU type "Analogous") and, in addition, the value does not have a structured data type. An *Alarms and Events Server* does not send any values to the client, but sends information that something has happened (e.g. a temperature has exceeded a limit value). The criteria the *Alarms and Events Server* uses for its decision that an event has occurred has to be set (e.g. limit value = 40 °C). This definition has to be carried out during adaptation of the server to a concrete application for the variables and devices to be monitored. The specification does not prescribe how to carry out the decision or how to indicate criteria. It is possible to define decision criteria for any variables (analogous, binary, structural data type, simple data type) and even a linkage of variables (temperature and time).

Alarms and Events Servers can be implemented as individual components that can obtain information directly from devices and applications or from *Data Access Servers*. In the last case, they need to have parts of a *Data Access Client* implemented. An *Alarms and Events Server* can also be implemented in a component together with a *Data Access Server.*

Alarms and Events Clients can be integral parts of control stations, recording components or management systems. Relatively simple components (Excel-Sheet) and complex programs, e.g. process control systems are possible.

The *Alarms and Events Specification* also supports the fact that a client can have access to several servers and one server can be used by several clients.

The specification defines an *event* as a detectable occurrence (e.g. failure of a device, excess of a limit value, operator intervention). Several events can be allocated to specific variables. For one temperature, four limit values can be defined, for example. This means that four events may occur, every one of them indicating that an upper or lower limit value has been exceeded. For implementing this model, the *Alarms and Events Specification* defines the concept of conditions and subconditions allocated to it. An event occurs if a condition or a subcondition allocated to it becomes active, e.g. the temperature has exceeded an upper or lower limit value. Thus the event signals an alarm, which occurs as long as the subcondition is active.

The specification defines the following event types which correspond to the model:

- Condition-related Events. Sending the event to the client shows that an alarm has occurred. Conditions can be activated/deactivated, i.e. their evaluation is switched on or off in the *Alarms and Events Server*. In this way, it can be determined whether events are monitored or not. There is an acknowledging mechanism for events which have occurred.
- Tracking-related Events. These provide information on the occurrence of an event the cause of which is not within the process (e.g. operator intervention) and which has to be tracked.
- Simple Events. They can be used to notify the occurrence of simple events (e.g. failure of a device) to the client.

Table 2.10: *Functionality for condition-related Events*

Functionality	**Meaning**
Existence of sub-conditions	Sub-conditions can be defined for a condition-related event. These can describe one (opening of a door) or several causes of an alarm (exceeding an upper or lower limit value).
Deactivation/Activation of monitoring	Evaluation of the condition in the server can be activated or deactivated
Acknowledging of an event	The server can request the client to acknowledge an event. The client then calls a method of the server
Connection of the event with an item in the Data Access Server	In order to be able to create an OPCItem Object, the client needs the ProgID of the Data Access Server, the fully qualified ItemID and possibly also the name of the computer on which the server runs. The Alarms and Event Server provides functionality which a client can use to get this information for an event. The client transmits the value of an event attribute (generally the complete identifier from the eventarea) to the method and receives the information in the form of return values.

The specification not only defines an event model and possibilities for transferring of events to the client, it also defines functionality which can be used with condition-related events. **Table 2.10** describes these possibilities. This functionality is not available for the other two event types.

The specification defines two ways to structure events:

- The eventarea is used for topographic organization of events and can be compared with the namespace of the *Data Access Specification*. The nodes of the eventarea describe areas in which events may occur (e.g. a room). The leaves describe event sources. Several events can be available at one event source (e.g. monitoring of the temperature value and the temperature sensor).
- The filter space is used for the logical structuring of the events. The specification distinguishes three event types, and there are event categories for these. Some

Table 2.11: Event categories and conditions

Event type	Event category	Condition
Condition related	Level	PVLEVEL
		SPLEVEL
		LO_LO
		LO
	Deviation	DV_LO
		DV_HI
	Discrete	CFN
		TRIP
Simple	Device Failure	
	Batch Status	
	System Message	
Tracking	Operator Process Change	
	System Configuration	

categories are already defined in the specification. For condition-related events, the existence of condition and subconditions for event categories is still defined. **Table 2.11** shows suggestions from the specification that can and probably will be enlarged in real applications. A severity level can also be assigned to an event. The filter space results from the attribute values assigned to events during configuration. A server will implement means of monitoring different events. The events can be assigned to event sources in the configuration. A fully flexible server that allows defining the event recognition also during configuration is possible, but its implementation involves much work.

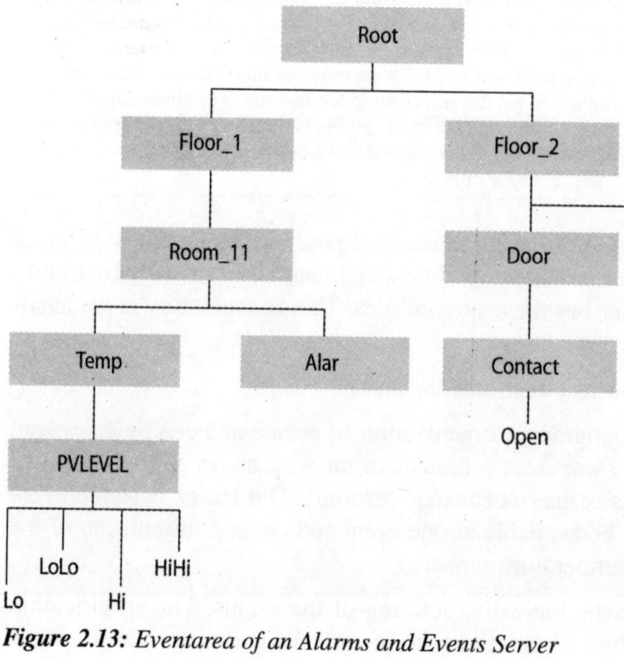

Figure 2.13: Eventarea of an Alarms and Events Server

Figure 2.13 shows an example for an eventarea in building automation. There is a (virtual) root. The next nodes mark two floors. For the node "Floor_1" the node "Room_11" is also available. Two event sources exist there – "Temp" and "Alarm." "Temp" is a condition-related event with the condition "PVLEVEL" and corresponding subconditions. "Alarm" is a simple event. For the node "Floor_2" there are two event sources – the condition-related event "Door" with the condition "Contact" and the subcondition "Open" as well as the "Temperature" setting as tracking-related event. The type of event is not indicated in the eventarea. The information is in the filter space.

The filter space marks the structure of the events according to their attributes (type, category, severity, etc.). One example is shown in **Figure 2.14**.

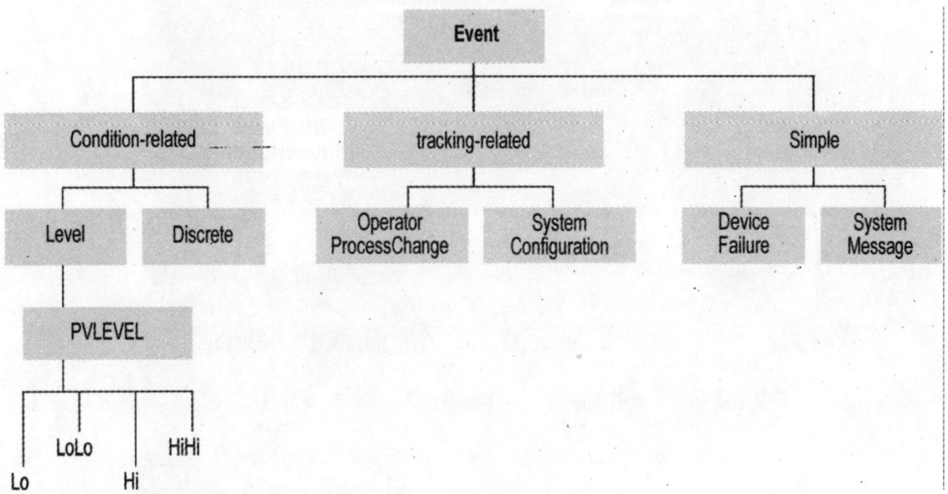

Figure 2.14: *Filter space of an Alarms and Events Server*

An *Alarms and Events Client* can create different OPC objects in the *Alarms and Events Server*. **Figure 2.15** shows the hierarchy of an example. The *OPCEventServer* object represents the top level object in the hierarchy of OPC objects. The optional *OPCEventAreaBrowser* object is used for browsing the eventarea. This differs from the *Data Access Specification.* There, the browser functionality is available at an interface at the OPCServer object. *OPCEventSubscription objects* form the lowest level. There are no explicit objects available for the events. This is a further structural difference compared to the object hierarchy in the *Data Access Server*; OPCItem objects are available there. Allocation of the events to the OPCEventSubscription objects is made via filters. The filter can contain value sets of event attributes or value sets of complete identifiers from the eventarea. For the *Alarms and Events Server* mentioned in **Figure 2.15**, allocation of events to the OPCEventSubscription objects could be effected either by using a filter criterion from the eventarea (events in "Area_11") or by taking it from the filter space (condition-related events with severity > 800).

For an *Alarms and Events Server,* the number of hierarchy levels for the OPC objects is determined as well; however, no restrictions are made as to the eventarea. For the filter space, only the number of hierarchy levels is determined. However, freely accessible event categories, conditions, and subconditions can be added. For the severity category, a range (1–1000) and the significance (high–low severity) of sections in the range have been determined. Priorities from real applications must be mapped to this range.

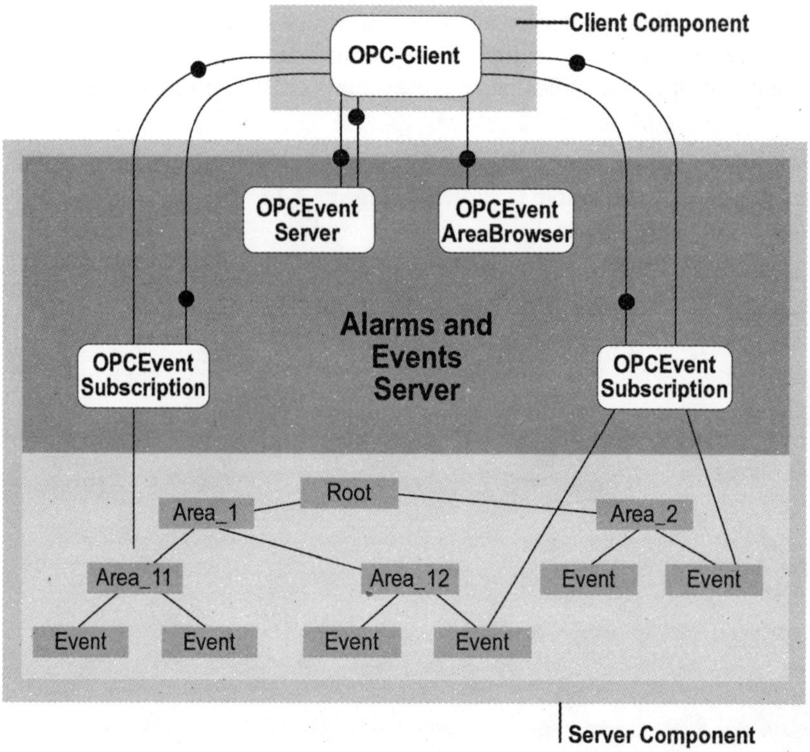

Figure 2.15: Alarms and Events Server – Object hierarchy and Eventarea

There is a callback connection from the OPCEventServer object to the client. This is used by the server to inform the corresponding client of the shutdown. Further callback connections are available at the OPCEventSubscription objects. They are used to transfer the event notification.

2.9.2 Specification object and use

Some of the specification objects have already been presented in the previous section – object hierarchy, eventarea, and filter space. Besides these issues, the definition of the format of the event message from the server to the client was made in the specification.

If an event occurs, the server sends the corresponding notification to the client via the callback connection. The notifications have various mandatory attributes for the different event types. Optional attributes can be defined in the application. They are valid for specific event categories. **Figure 2.16** shows the mandatory attributes. The attribute values mentioned correspond to the notification of a condition-related event.

If a simple event occurs, the server always has to transfer the description of the source, the time at which the event happened, the type of the event, the event category (Category), the severity, and a message. Since the *IOPCCommon* interface is also available in an *Alarms and Events Server,* the client can query whether the server supports additional languages and can request that the message is passed in the desired language.

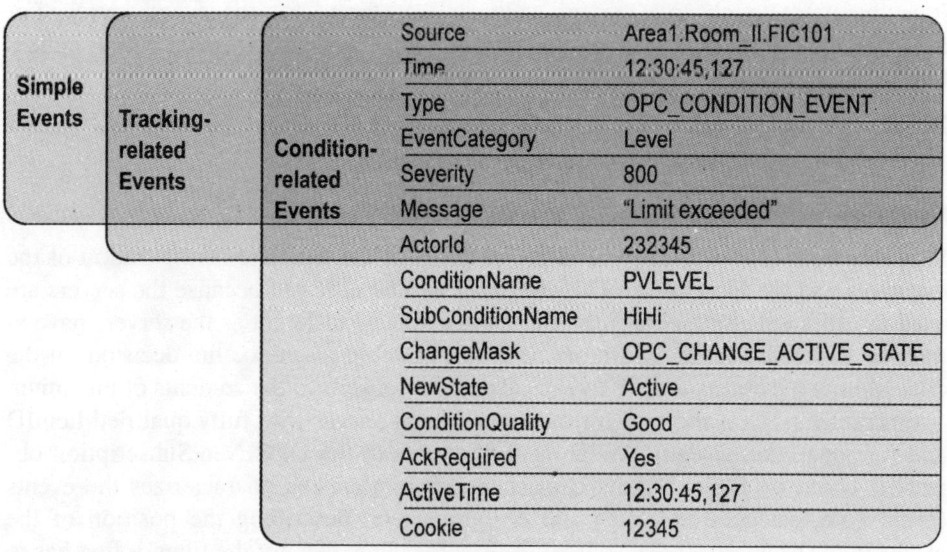

Figure 2.16: *Attributes of the event notifications for different event types*

For a tracking-related event, the ActorId is added to these six mandatory attributes. It marks the origin of the event. Concrete parameter value depends on the application.

The largest number of attributes has to be transferred for a condition-related event. Acknowledging such an event leads to another event as the status of the condition/ subcondition has changed. The ActorId is here used to mark the origin of the acknowledgment. Besides the attributes already mentioned, the following are added – the name of the condition (ConditionName) and, if available, the name of the subcondition (SubConditionName). The value of the attribute ChangeMask indicates which state change is communicated by the event notification. The new state is also transferred. The values in **Figure 2.16** show that the subcondition "HiHi" of the condition "PVLEVEL" has the actual state "Active." Comparable to the status in the data format of the *Data Access Server*, a ConditionQuality also exists in the *Alarm and Event Server.* It gives information on the status of the variable linked with the *Condition.* The values this attribute can take are the same as the values of the "Quality" with the *Data Access Server.* The status can thus be "Good," "Bad," or "Uncertain." With the *Data Access Server*, a notification is automatically sent to the Client if the status is changed for active OPCItem objects in active OPCGroup objects. The same is true for the *Alarms and Events Server.* If the status changes from "Good" to "Bad" – for instance, if the process variable can no longer be accessed – a notification is sent to the client. The value of the parameter AckRequired determines whether the event is to be acknowledged or not. The ActiveTime indicates when the condition has become "Active." The value is identical with the value of the attribute Time when notification has been sent to the client of the event linked to the activation of a condition. These values differ if the event shows other modifications of the state of the subcondition, e.g. that an acknowledgment has taken place; the new state is then acknowledged.

It is important to observe that no value is transferred for the event source. For example, the information is given that a limit value has been exceeded but no indication of the current temperature.

Below, the functionality of the current version (1.10) of the *Alarms and Events Specification* is described using specific scenarios.

Querying for filter criteria

Real *Alarms and Events Servers* differ in terms of the concrete configuration of the eventarea and the filter space. The eventarea will be different because the servers are used for different applications; the filter space will be different as the servers have to monitor different events. Therefore, it is not possible to reduce the decision on the allocation of the events to OPCEventSubscription objects to the contents of maximum two parameters, as is the case for the *Data Access Server* with fully qualified ItemID and (optional) AccessPath. Assignment of events to the OPCEventSubscription objects is based on filters. Filters consist of information that characterizes the events (event type, category, severity) and of information describing the position of the events in the eventarea (area, source). Before the client can use the filter, it first has to query for the filter criteria available. A server can support filtering based on the following values:

- values for the event type (condition related, simple, tracking related)
- values for the event category
- values for the severity
- values for the area (nodes in the eventarea)
- values for the sources (leaves in the eventarea)

For the possibilities supported by the server, the client will then request the values available in the next step.

Different event categories are available for the three event types. With condition-related events, different conditions are defined for the various event categories. These conditions on their part can have subconditions. The whole amount of conditions and subconditions is not directly used for filtering. The client can call a method which returns conditions for sources in the eventarea. These event sources form an integral part of the filter. It's advantageous in this procedure for the client to use defined categories (PVLEVEL, etc.), conditions, and subconditions to request the event sources available.

After finishing this step, the client has a large variety of values with which it can create filters.

Browsing of the eventarea

An *Alarms and Events Server* can make available an eventarea in which all the events are structured. The specification assumes that the eventarea is always hierarchical and is structured into areas and sources.

By using the method *ChangeBrowsePosition*, the client is able to move to specific nodes (areas) in the eventarea of the *Alarms and Events Server*. When calling the method *BrowseOPCAreas*, a list of nodes (areas) or leaves (sources) is returned. The amount of return values can be limited by textual filters (T*).

The two methods *GetQualifiedAreaName* and *GetQualifiedSourceName* return fully qualified SourceId or AreaId which can be used as filter criteria. It becomes clear that nodes have an enlarged semantic significance in the eventarea compared to nodes in the namespace of the *Data Access Server*. Node identifiers can be used as filter parameters or for defining areas for the deactivation/activation of conditions.

Query for optional attributes for the event notification

In the event notification, a whole series of mandatory attributes is passed to the client. These attributes are sometimes not sufficient to transfer all the important information to the client. When adapting the server to a concrete application, additional attributes can be assigned to the different event categories. One example would be a pointer to a set of instructions to be carried out if an event occurs. First of all, the client is able to query for additional attributes for an event category in order to determine later, after finishing the assignment of events to the OPCEventSubscription objects, which of these optional attributes are to be passed in the notification besides the mandatory attributes.

Creation of OPCEventSubscription objects

OPCEventSubscription objects are used to monitor the event sources in the server and inform the client of the events which have occurred. The client is able to create an object every time the corresponding method is called. The filter for a newly created object is empty at first, which means that the object is responsible for monitoring all events in the server from the time it has been created.

When creating the objects, the client transfers values for the parameter *Active*, *MaxSize,* and *BufferTime*, etc. These values influence the behavior of the object and the reaction time to be achieved.

If the object is in an inactive state, no monitoring of the events allocated to the object is carried out and therefore no event notifications are sent to the client. As a newly created object is responsible for monitoring of all events, it should always be created "inactive" as otherwise notifications could be sent for events. In the "active" state, the frequency with which notifications on events occurred are sent is determined by the parameter values *BufferTime* and *MaxSize*.

Events are transferred to the client by means of callbacks. If every event is transferred individually, resources are needed for the transfer and the processing in the client is constantly interrupted. If several events are combined, transfer is made more efficiently. On the other hand, the client should be informed as quickly as possible of events. These two requirements can be met when *BufferTime* and *MaxSize* are set respectively. With the *BufferTime* value (set in ms) the client determines the time interval at which the events are to be sent. A value of 100 ms determines that events that have occurred are sent every 100 ms. All events that have occurred before that time elapses are buffered until then.

Should many events occur simultaneously, which would indicate a number of exceptions occurring in the process, their notification to the client should be carried out much more quickly. This can be influenced by the value of *MaxSize* which determines the length of the event queue in the server in relation to the OPCEventSubscription object. A value of 5 would then determine that a notification is sent to the client as soon as 5 events have arrived. This is done independently of whether the *BufferTime* has run out, and can happen after 50 ms for the example above.

Which is the significance of the value 0 with these two parameters? A *BufferTime* of 0 determines that every event is transferred immediately after it has occurred. A *MaxSize* of 0 means that the queue can be of any size and that events are transferred to the Client when the *BufferTime* has run out.

Use of the OPCEventSubscription objects

The method *SetFilter* enables the client to assign events to the OPCEventSubscription object by using filters. The client has already queried for the available filter values. A filter is composed of different criteria. Values for one criterion are logically OR'ed together, and values of different criteria are logically AND'ed together. An OPCEventSubscription object can monitor events of either simple or condition-related type which have a severity of more than 700. The client is able to request the current filter by means of the method *GetFilter*. **Table 2.12** shows the filter criteria and some values. Furthermore, the table contains comments on specific values for filter criteria.

Table 2.12: *Values for filter criteria*

Filter criteria	Values	Annotation
EventType	(condition-related)	A value of 0 is an error
Category	(Level, Deviation)	A NULL pointer indicates that all categories have to be included
LowSeverity	(100)	A value of 0 indicates that events with all severities are to be received, even if HighSeverity is 1000
HighSeverity	(400)	A value of 1000 indicates that events are to be received independent of their severity, even if LowSeverity is 0
AreaList	(Area_11, Area_12)	A NULL pointer indicates that all the areas are to be included
SourceList	(Source_1, Source_2)	A NULL pointer indicates that all the sources are to be included. Wild cards (* or ? etc.) are also possible.

Besides the mandatory attributes for the event notification, the client can also request the addition of further optional attributes. This can be done using the method *SelectReturnedAttributes*. The client has already queried for the attributes possible for an event category. The telephone number of a service engineer could be transferred in such an optional attribute. The user at the client would then know whom to inform in case of an event. By using the method *GetReturnedAttributes*, the client can query for the optional attributes which have to be sent with the event notification.

When calling the method *Refresh*, the client gets information on all active and inactive, non-acknowledged conditions. The method can only be applied to condition-related events. A client could carry out a refresh call after it has created its OPC object hierarchy in an already running *Alarms and Events Server*, in order to get a current overview of all condition-related events (alarm states). The server will then send event notifications for all active and inactive, non-acknowledged events, which correspond to the filter of the subscription object, to the client after receiving the refresh call. Please consider that the overview only contains the event last occurred. A functionality, in the sense of an event history, is not available. An outstanding refresh can be aborted by the client using a *CancelRefresh*.

OPCEventSubscription objects are always private. A concept comparable to that of the public OPCGroup objects in the *Data Access Server* does not exist.

Acknowledging and deactivation/activation of events

The methods *EnableConditionbyArea/EnableConditionbySource* and *DisableConditionbyArea/DisableConditionbySource* are used to switch on or off monitoring of condition-related events (alarms) for an area or a source of the eventarea. If, for example, maintenance work is to be carried out in parts of the plant, it is not necessary to monitor all events.

The method *AckCondition* is used by the *Alarms and Events Client* to acknowledge condition-related events that have occurred, if this is required in the corresponding attribute of the event notification. A new event occurs if the acknowledgment arrives in the server. This makes it possible to keep an alarm history on the client's side. There, the following can be recorded:

- When the alarm occurred (value of the attribute *Time* of the first notification or value of the attribute *ActiveTime* of the second notification),
- When the acknowledgment was made (values of the attribute *Time* of the second notification)
- A hint to the person who acknowledged the alarm (a value of the attribute *ActorId* of the second notification)

The method *GetConditionState* can be used by the client to request the current status information of a condition. The following information is passed to the client:

- Condition state
- Existing subconditions, and information about them
- Severity
- Time at which the last event occurred or when it was acknowledged

The differences between the methods *GetConditionState* and *Refresh* are that they are implemented at different objects and that the conditions are addressed differently.

A *Refresh* relates to condition-related events managed by an OPCEventSubscription object. Addressing is made implicitly, i.e. all condition-related events of this object are concerned. Event notifications are returned for active and inactive, non-acknowledged events. With the *GetConditionState*, the client is able to request specific information on conditions available in a source of the eventarea. Similar to *Refresh*, *GetCondtionState* could be invoked by a client when starting a server in order to get information on the state of the various conditions.

Status information about the OPCEventServer object and the OPCEventSubscription object

The method *GetStatus* is used by the client to get information on the status of the server.

The results passed to the client are similar to those obtained when the *GetStatus* method is used on the *Data Access Server*, i.e. information about vendor name, start time, and version of the server. Unlike the *Data Access Server,* where the client was able to ask for the number of OPCGroup objects, the number of OPCEventSubscription objects is not part of the response. The methods *GetState* and *SetState* enable the client to influence the state of the OPCEventSubscription object. The client is able to set here the state (active/inactive) and the values for the parameters *BufferTime* and *MaxSize*.

Link to the Data Access Server

Alarms and Events Servers can be used in connection with *Data Access Servers*. Both servers make information available that is derived from the same data sources – e.g. the current value of a temperature and the event notification when exceeding the limit value of the temperature. The method *TranslateToItemId* enables a client to query for a relationship between an attribute of the event notification and an item in the namespace of the *Data Access Server*. This optional method returns the information for event attributes (e.g. source) of the condition-related events necessary for a *Data Access Client* to create an OPCItem object. The information consists of a triple <NodeName, CLSID, fully qualified ItemId>. The specification does not describe how this information is exchanged between the *Alarms and Events Client* and the *Data Access Client*.

Figure 2.17 shows the objects of an *Alarms and Events Server* with their interfaces. The client object with the callback interfaces is also included.

The OPCEventServer object is used as a logical container of the OPCEventSubscription objects. This object supports the following three interfaces: *IOPCCommon* (see section 2.2), *IConnectionPointContainer,* and *IOPCEventServer.* The functionality defined for this object in the specification can be used by calling methods at the interface *IOPCEventServer.*

The object *OPCEventAreaBrowser* has an interface *IOPCEventAreaBrowser.*

The object OPCEventSubscription has two interfaces. At the interface *IOPCEventSubscriptionMgt*, methods are available which enable access to the functionality of the object (creation and request of filters, etc.) described in the specification.

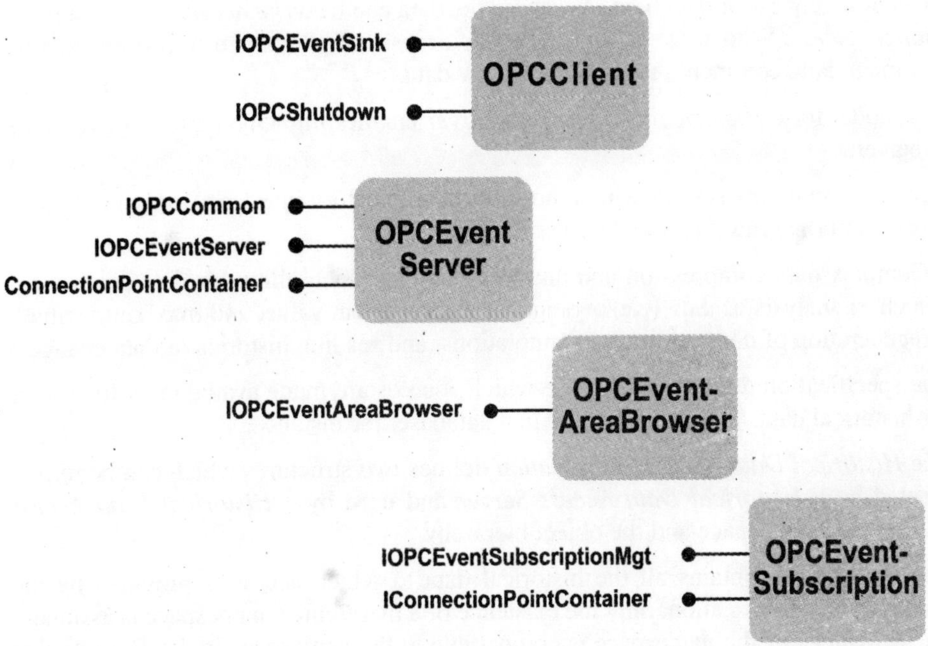

***Figure 2.17:** OPC objects in the Alarms and Events Server and Client*

An *Alarms and Events Client* makes available two callback interfaces – *IOPCEventSink* for transfer of the event notification and *IOPCShutdown* for transfer of the shutdown information.

The two interfaces *IOPCEventServer2* and *IOPCEventSubscriptionMgt2* were added in version 1.10. *IOPCEventServer2* extends the functionality with regard to activating and deactivating event monitoring. *IOPCEventSubscriptionMgt2* allows connection monitoring for subscriptions.

Automation Interface

Specification of the automation objects for the *Alarms and Events Server* was started, but has not been finished. As work on this specification is not likely to be completed, the current status of the specification will not be described. OPC Alarms and Events servers can also be accessed from languages using automation controllers by means of ActiveX controls. An example is shown in section 3.5.4.

2.10 OPC Historical Data Access Specification

2.10.1 Field of application and model

The explanations in this section relate to version 1.20 of the named specification.

Historical Data Access Servers enable clients to access historical data. The two basic types of data are *raw data*, which means the historical data stored, and *aggregated data*, which means data extracted from the raw data. These aggregates are only created on request of the client. Access to the data can be *write access, read access* or *change access*. With write access or change access, the client can overwrite existing historical data, comment them, or insert new data.

The model for a *Historical Data Access Server* enables implementation of two kinds of servers:

- Simple trend data servers which, besides the mandatory interfaces, only implement some optional interfaces and access to raw data.
- Complex data compression and analysis servers, which allow for data processing, such as analysis of data (e.g. average value, minimum value, and maximum value), regeneration of data, addition of annotations, and reading histories of data changes.

The specification does not determine which sources are made available by the server for historical data. The data could be in a database, for instance.

The *Historical Data Access Specification* defines two structures which can be implemented by a *Historical Data Access Server* and used by a *Historical Data Access Client* – the namespace and the object hierarchy.

The namespace contains all the historical data to which access is provided by the server. In the specification, only the existence of a hierarchical namespace is assumed. The semantics of the namespace is comparable to the namespace in the *Data Access Server.*

A *Historical Data Access Client* can create different OPC objects in the *Historical Data Access Server.* The *OPCHDAServer Object* is the top-level object. It provides the complete functionality for reading, writing, or altering the data and their attributes. The *OPCHDABrowser Object* offers functionality for browsing the namespace.

Unlike the *Data Access Server,* there are no objects comparable to the OPCGroup and OPCItem objects. The client directly addresses data items by means of the handle without creating objects for this purpose in the server. Why was this model chosen?

A *Data Access Server* allows permanent access to process variables which can be structured according to different aspects. Permanent adding and removing of OPCItem objects for the same item in the namespace is rather an exception. The number of process variables made available by a *Data Access Server* is in the range of 100 – 1000 or above. The client wants to read or write exactly one value of a process variable. This is different with a *Historical Data Access Server.* The number of process variables available is increased in most of the cases by a factor of 10 (1000 – 10000). The client, however, is not willing to read these variables constantly, but probably only once a day. Instead, the client will not only be interested in one single value but in several values (e.g. a history), or in an aggregate (e.g. average value). A large flexibility in defining these aggregates could, of course, become a distinguishing feature for the *Historical Data Access Server.*

Figure 2.18 shows an object hierarchy and the relation to the namespace.

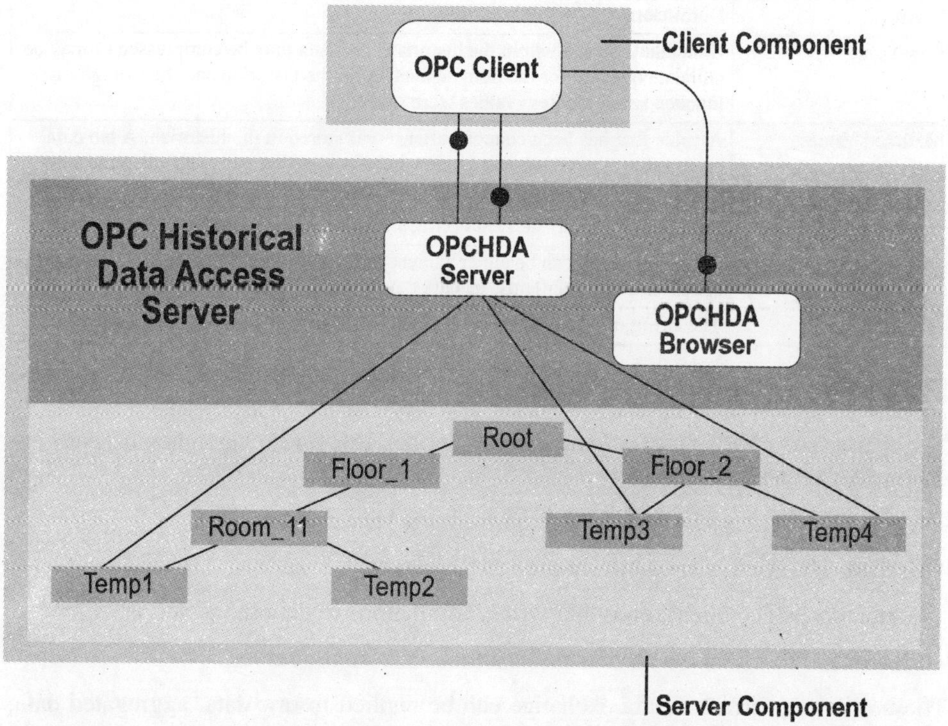

Figure 2.18: *Historical Data Access Server – Namespace and object hierarchy*

As a basis for structuring the data and creating aggregates, the specification defines the concepts mentioned in **Table 2.13**.

Table 2.13: *Concepts for structuring of data in the Historical Data Access Server*

Term	Definition
Attributes	An additional qualifier that a particular item may have associated with it. For example, an "Item Value" property would probably have the following attributes associated with it: "Data Type" (VT_R4), "Stepped" (0) and "Archiving"(1)., i.e. the "Item Value" returns a 4 Byte real number, the value can be displayed as interpolated (sloped line) and data is being archived. This corresponds to the properties from the Data Access Specification.
Aggregates	Methods that summarize data values. Common aggregates include Aver ages over a given time range, Minimum over a time range and Maximum over a time range. These aggregates are performed during the retrieval of the data.
Annotation	An operator or user entered comment that is associated with an item, usually at a given instance in time. There does not have to be a value stored at that time.
StartTime/End Time	Times which limit a request. The times belong to the request (31.07.12:00:00 31.08.12:00:00)
Bounding Values	Bounding values are required by clients to determine the entry and exit points when requesting raw data over a time range. If a raw data value exists at the entry or exit point, it is considered the bounding value even though it is part of the data request. If no raw data value exists at the entry or exit point, the next data point outside of the range is considered the bounding point.

Term	Definition
Raw Data	Data that is stored within the historian. The data may be compressed or may be all data collected for the item depending on the historian and the storage rules invoked when the item values were saved.
Modified Values	A value that has been changed after it was stored in the historian. A lab data entry value is not a modified value, but if a user corrects a lab value the new value would be modified.
Interpolated Data	Data that is linearly derived between two stored data points.
ItemHandles	The ItemHandle can be either a client or server value. It is used by the owner to speed access to the items. Its data type is OPCHandle (DWORD).
ItemId	The string that is a reference to a piece of data in the server address space.

2.10.2 Specification object and use

The *Historical Data Access Specification* defines the following types of data exchange between client and server:

- Read. The client reads the information from the server.
- Update. The client writes information to the server.
- Annotations. The client reads and writes annotations to data in the server.
- Playback. The client gets recurrent information from the server.

These different forms of data exchange can be applied to raw data, aggregated data, modified data, and attributes of data.

Table 2.14 gives an overview of the different possibilities of data exchange between *Historical Data Access Clients* and *Historical Data Access Servers*:

Table 2.14: *Possibilities of data exchange in Historical Data Access Servers*

Type	Kind of information	Annotation
Read		
	Raw	Reading of value, quality and time stamp for the indi cated items within a predetermined period of time
	Processed	Reading of aggregated values, quality and time stamp for the indicated items within the predetermined period of time
	AtTime	Reading of value, quality for the indicated items at a predetermined period of time
	Modified	Reading of value, quality and time stamp for the indi cated items within a predetermined period of time. The values have been modified or replaced by other functions
	Attribute	Reading of the attributes of the items for a specified period of time
Update		
	Insert	Inserting of values and quality at specific times
	Replace	Replacing of values and quality at specific times

Type	Kind of information	Annotation
	InsertReplace	Inserting and replacing of values and quality at specific times. If a value already exists at this time, it will be replaced, otherwise inserted
	DeleteRaw	Deleting of value, quality and time stamp for the specified period of time
	DeleteAtTime	Deleting of value, quality and time stamp for the specified period of time
Annotation		
	Read	Reading of annotations for the indicated period of time
	Insert	Inserting of annotations for the indicated period of time
Playback		
	ReadRawWithUpdate	In defined intervals, the raw values are played back for a defined period of time. This can be used for displaying a trend.
	ReadProcessedWithUpdate	In defined intervals, the processed values are played back for a defined period of time. This can be used for displaying a trend.

Browsing of the namespace, request and release of ItemHandles

The namespace of the *Historical Data Access Server* has the same semantic significance as the namespace with the *Data Access Server* – it is used to structure the existing data and offer them to the client. With the *Historical Data Access Server*, access to the namespace takes place by invoking methods at an interface of the OPCHDABrowser object. After finishing namespace browsing, the client disposes of a number of fully qualified ItemIDs. The procedure can be compared to browsing the namespace in the *Data Access Server.*

After browsing the namespace, the client will call the method *GetItemHandles* with the fully qualified ItemIds returned in order to get handles from the server. This enables the client to address process variables in the server. By using these handles, data exchange will take place later. If the client no longer wants to carry out any access to the items, it calls *ReleaseItemHandles*. The server recognizes that the client no longer wants to carry out any access with the ServerHandle; the server can then use the handle for other purposes.

Reading of values (Read)

The OPCHDAServer object implements synchronous and (at an optional interface) asynchronous method calls for reading the values. The process with asynchronous reading is analogous to the process of asynchronous calls in the *Data Access Server.*

What different kinds of historical data values can be read by the client?

- Raw data for a period with or without values for start and end time, e.g. from 07/31/00 12 am to 08/31/00 12 am. Not only values but also the quality and the time stamp are returned.
- Processed raw data for a period. A *Historical Data Access Server* can make available different aggregates for processed raw data. The client can query for existing

aggregates. A number of aggregate types have already been defined in the specification (e.g. average value, minimum, maximum, time average). Further types can be implemented.

- Values at a fixed time. By using this method, values can be read that correlate with other values at the same time (e.g. all the temperature values at 5 pm).
- Modified and deleted values. By using this method, the client is able to read the initial values of deleted or modified values.
- Reading of item attributes for a specified time period. A server can support a number of attributes for items. Some of the attributes are already defined in the specification (e.g.) VARIANT data type, description, reference to origin of the data). Furthermore, manufacturer-specific attributes can be implemented. This corresponds to the properties from the Data Access Specification.

Figure 2.19 shows the relation between raw data, aggregates, and attributes. The sector of the database for a *Historical Data Access Server,* which is explained in the following, is represented there.

Raw Data										
	1	2	3	4	5	6	7	8	9	10
Temp1	18	18	19	20	21	20	19	18	17	18
Temp2	17	17	18	19	20	17	17	17	18	18
Temp3	16	17	18	18	18	17	18	19	19	18
Temp4	17	18	18	19	20	20	20	20	19	19

Attributes				
	Temp1	Temp2	Temp3	Temp4
Datatype	VT_I2	VT_I2	VT_I2	VT_I2
Description	Temperature in Room11	Temperature in Room12	Temperature in Room13	Temperature in Room14
NodeName	128.7.16.123	128.7.16.123	128.7.16.124	128.7.16.124

Aggregates				
	Temp1	Temp2	Temp3	Temp4
Average	18,8	17,8	17,8	19
Delta	0	1	2	2
Minimum	17	17	16	17
Maximum	21	20	19	20
Range	4	3	3	3

Figure 2.19: *Relation between raw data, attributes and aggregates in a Historical Data Access Server*

Four temperature values (Temp1–Temp4) have been stored in a database. The figure shows values for 10 timestamps. Values exist for all timestamps. The items also have attributes. Example values are given for three attributes. The attribute NodeName designates the computer – on which an OPC server is running – that has supplied the value. The results of five aggregates are represented for the items. These values are always created according to the current call. The significance of the aggregates is described in **Table 2.15**.

Table 2.15: *Significance of aggregates in a Historical Data Access Server*

Aggregate	Meaning
Average	Retrieve the average data over the resample interval.
Delta	Retrieve the difference between the first and last value in the resample interval.
Minimum	Retrieve the minimum value in the resample interval.
Maximum	Retrieve the maximum value in the resample interval.
Range	Retrieve the difference between the minimum and maximum value over the sample interval.

Update of values

The OPCHDAServer object implements synchronous and asynchronous methods for the update of values. The synchronous and the asynchronous interface are both optional. Furthermore, additional optional methods exist at the interfaces.

Which possibilities are available for the client to insert data?

- Insert values in the database. It is possible to insert values for the variable, the time stamp, and the quality. Insertion is not possible if information for the item at the time indicated is already available in the data base. A corresponding error value is then returned.
- Replace values in the database. Values can be replaced. However, replacing is not possible if no value is available in the data base for the item at the time indicated. A corresponding error value is then returned.
- Insert/replace values in the database. It is possible to insert or replace values for the variable value, the time stamp, and the quality. The information is inserted if no value is available in the database for the item at the time indicated. If there are already values available in the database, they are overwritten.
- Delete raw data from the database. For a certain time range, values of one or more items can be deleted from the database.
- Delete data at time. All the values for the items indicated are deleted for specified timestamps.

Annotations

The OPCHDAServer object implements synchronous and asynchronous methods for access to annotations. The synchronous and the asynchronous interface are both optional.

The client has the following possibilities:

– Read annotations. The existing annotations for the items can be read for an interval indicated by the client.

– Insert annotations. Supplements to items are made for specified timestamps.

Playback

Only an asynchronous interface exists for this functionality. When first called, the client gets an initial quantity of data; it then gets new data for a specific time range at certain intervals. The data are transmitted to the client via a callback connection.

– Read raw data with update. First, the server sends values for addressed items of the time interval passed with the method call. The client determines the frequency and the amount of data to be sent subsequently by passing two other time parameters to the server. This makes it possible to transfer data in real time, more slowly, or (for existing data) more quickly than the recording speed.

– Reading of processed data with update. The functionality corresponds to that explained above. The only difference is that not raw but processed data are returned. The client can specify what kind of aggregate it is interested in.

Figure 2.20 represents the objects of a *Historical Data Access Server* with their interfaces. The client object with the callback interfaces is also included.

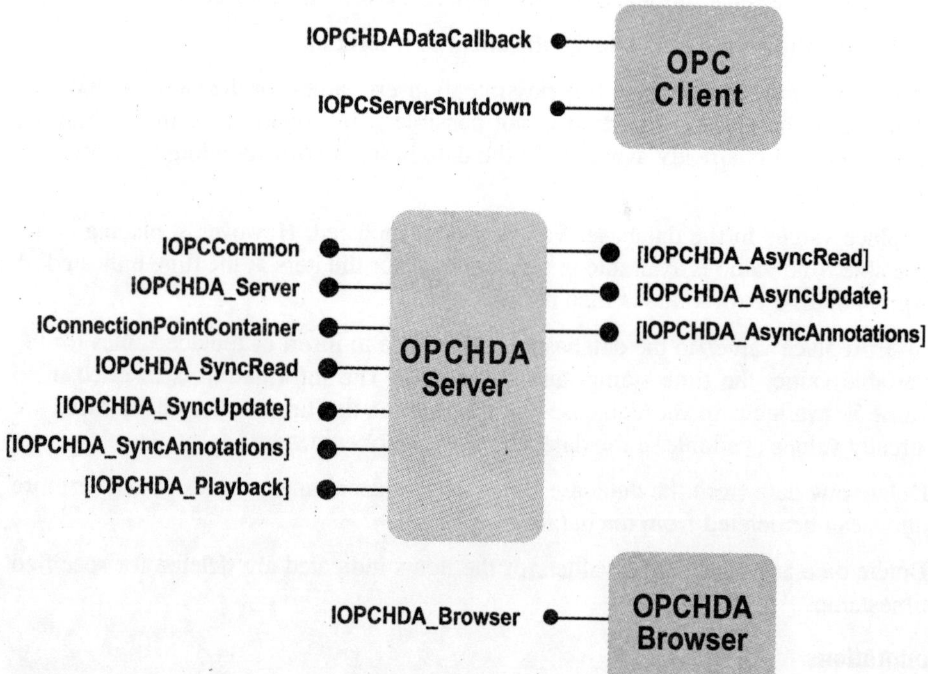

Figure 2.20: OPC objects in the Historical Data Access Server and Client

An HDA server implements only two objects.

All the interfaces with methods for access to historical data are located at the OPCHDAServer object. The object has four mandatory and six optional interfaces. At the interface *IOPCHDA_Server*, the functionality for requesting the supported aggregates and attributes is available. Synchronous and asynchronous interfaces are available for the different types of information request (Read, Update, Annotations).

The object OPCHDABrowser supports one interface.

A *Historical Data Access Client* makes available two callback interfaces – *IOPCHDA-DataCallback* for notification of the completion of asynchronous calls and *IOPCShutdown* for transfer of shutdown information.

Version 1.20 of the specification does not contain any new interfaces. It has mainly been extended by information clarifying the functionality and by further details on aggregates.

Automation Interface

The Automation Interface Specification for *OPC Historical Data Access Servers* exists in version 1.0. The OPC Foundation provides an implementation of the interface as a wrapper.

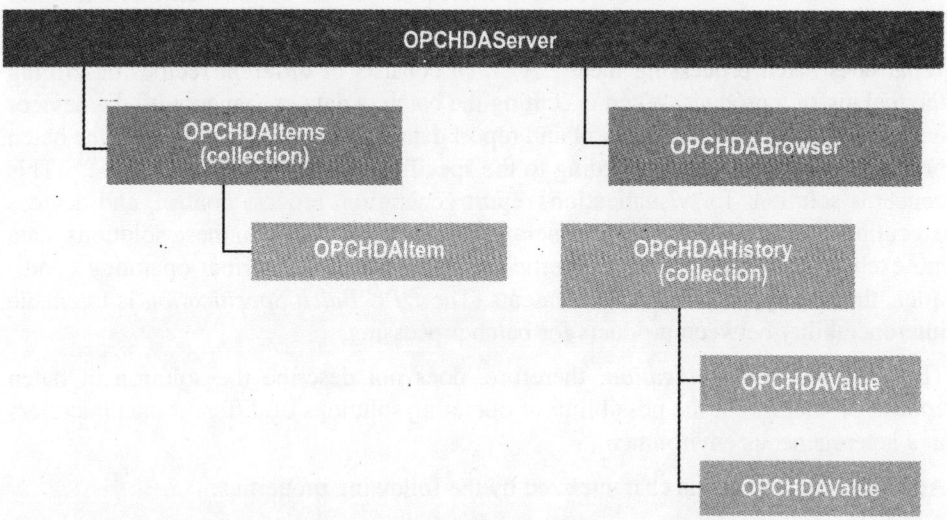

***Figure 2.21:** Automation objects of the Historical Data Access Specification*

Figure 2.21 shows the object model defined by the specification and implemented by the wrapper. In addition to the object model with custom interfaces, there are other objects, which are mentioned in **Table 2.16.**

***Table 2.16:** Additional automation objects for HDA Servers*

Object	Meaning
OPCHDAItems	A collection object that contains the OPCHDAItems created by a client. By using methods of this object, OPCHDAItem objects can be managed and their values can be read with the specified methods (see table 2.11).

Object	Meaning
OPCHDAItem	An object that represents an OPCHDAItem.
OPCHDAHistory	A collection object that contains either OPCHDAValue or OPCHDAEntry objects and represents part of their history. This object contains results of method calls used to access OPCHDAItem objects. For instance, the client can invoke a SyncRead for a specific OPCHDAItem object for a defined time span. The returned values can be assigned to an OPCHDAHistory collection.
OPCHDAValue	An object that represents a concrete value within the collection.
OPCHDAEntry	An object that contains attribute information on an OPCHDAValue object in the collection.

2.11 OPC Batch Specification

2.11.1 Field of application and model

The following information relates to version 2.0 of the *OPC Batch Specification.*

Unlike the specifications *Data Access, Alarms and Events,* and *Historical Data Access*, the *OPC Batch Specification* does not describe any completely new programming interface between a client and a server. It defines supplements to the *Data Access Specification* for the case of batch processing.

What does batch processing mean? A *batch* consists of different recipes describing the making of a product. When executing the batch, a data exchange with the devices is carried out. Recipe data are sent and report data are received. Products for the batch processing are developed according to the specifications of the IEC 61512-1[10] . This concerns solutions for visualization, report generation, process control, and devices executing control sequences and generating report data. Between these solutions, data are exchanged relating to the properties of the devices, the current operating conditions, the history, and the recipe contents. The *OPC Batch Specification* is to enable interoperability between products for batch processing.

The *OPC Batch Specification,* therefore, does not describe the solution of batch control problems, but the possibility of operating solutions of different manufacturers in a heterogeneous environment.

An *OPC Batch Server* is characterized by the following properties:

- Support of all mandatory interfaces and the optional browse interface of the *Data Access Specification* V2.0.
- Support of additional OPC interfaces, which are defined in the *OPC Batch Specification.*
- Existence of an exactly defined namespace, the structure and identifier of which are derived from the IEC 61512-1. Due to the large quantity of information created and exchanged with batch processing, the namespace can be very large.
- Values represented by items in the namespace of an *OPC Batch Server* have to be readable. They can be writeable as an option.

10 Standard IEC 61512 describes the aspects of batch processing.

OPC Batch Servers are based on the definitions of the batch processing standard IEC 61512-1. *OPC Batch Servers* enable clients to access batch information. The IEC standard distinguishes four types of batch information:

– Equipment capabilities

– Current operating conditions

– Historical contents

– Recipe contents

Batches are executed by devices which are described as equipment capabilities.

The object of version 2.0 of the *OPC Batch Specification* is the definition of a specific namespace and of interfaces for the exchange of information about equipment capabilities and current operating conditions between client and servers. Equipment capabilities correspond to the physical model of the IEC 61512. The current operating conditions are described in the IEC 61512 using a batch list and a batch model.

The IEC 61512-1 defines different models for structuring batch processing. For the explanation of version 2.0 of the *OPC Batch Specification,* only the physical model and the batch model are of interest here.

The physical model subdivides a company into different levels: company, site, area(s), process cell(s), unit(s), device module(s), and procedural control module(s).

A procedural control module is subdivided into:

– procedure

– unit procedure

– operation

– phase

This subdivision is also reflected in the batch model. There, execution of a batch is divided into units, recipes, operations, and phases.

The IEC 61512 allows for individual levels of the model to miss or for additional levels to be added in real applications. Information on missing levels or inserted levels is given in the namespace on properties.

The IEC 61512-1 of an *OPC Batch Server* defines a recipe consisting of 5 types of information:

– Header

– Formula

– Device requirement

– Procedure

– Other information

In the namespace these information categories are mapped on properties of the nodes and leaves, which represent Procedural Control Modules.

The *OPCBBatchModel* Branch is used in the specification in order to describe the execution of batches. It establishes the relationship between the physical model and the procedural model.

Figure 2.22 shows an example of a namespace of a *OPC Batch Server*.

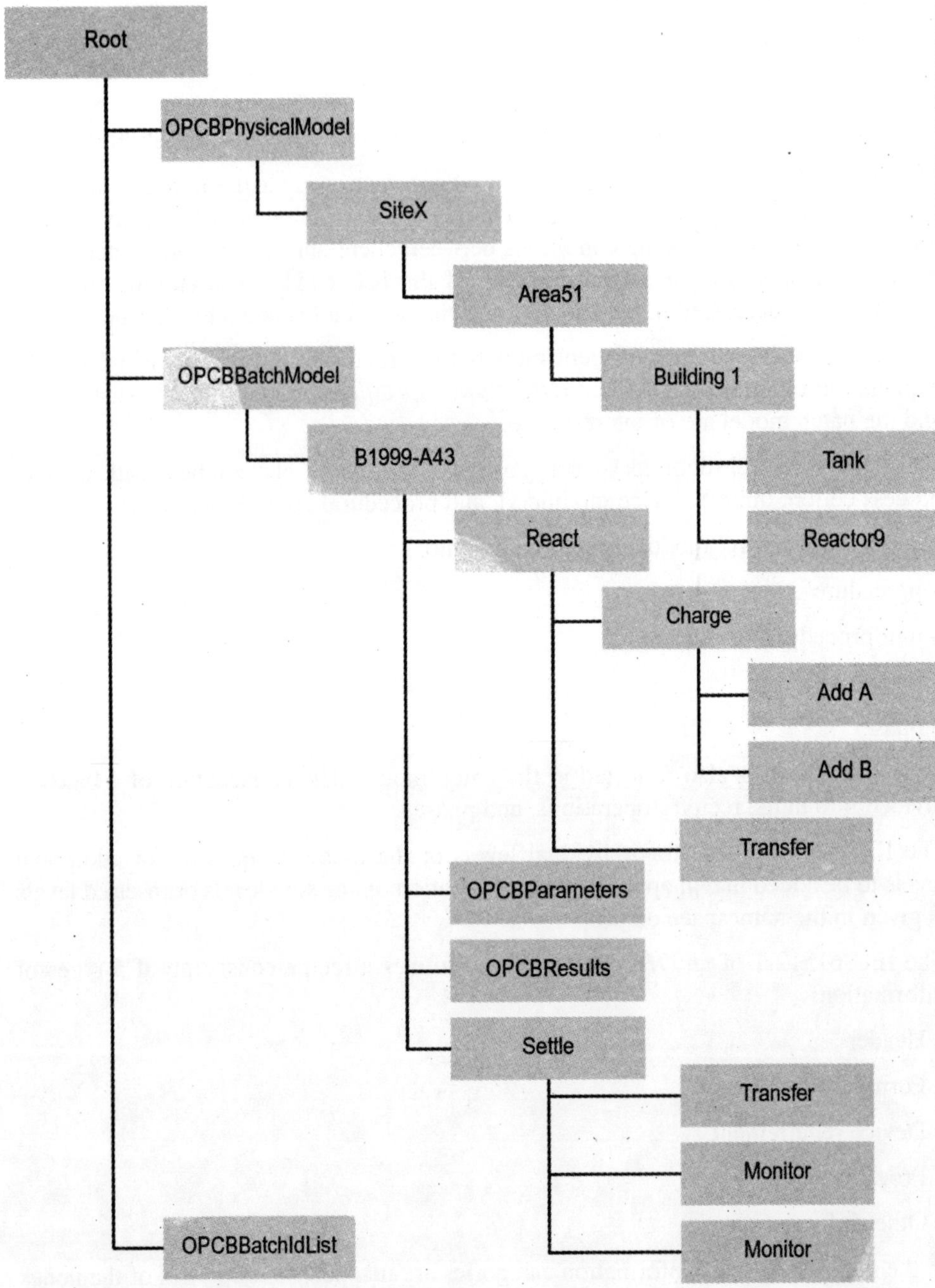

***Figure 2.22:** Namespace of a OPC Batch Server*

Some branches of the namespace represent models of the IEC 61512-1. They are located below the nodes OPCBPhysicalModel and OPCBBatchModel.

This *OPC Batch Server* enables access to processing in the whole company ("SiteX"). There, an "Area51" is available, in which "Building_1" is located with the processing units "Tank" and "Reactor9". In practical applications, further nodes may be available – devices and processing units in tank and reactor. All the information is contained in the *OPCBPhysicalModel* branch.

The *OPCBBatchModel* branch contains one batch – "B1999-A43". The batch contains two procedures – "React" and "Settle". These, in turn, contain operations with phases. Parameters and results can exist on every level in the namespace hierarchy in the *OPCBBatchModel* branch. Parameters are used to send information to the devices carrying out the operations, and the results are used to request values from the devices resulting from the execution.

2.11.2 Specification object and use

The functionality of an *OPC Batch Server* is to be explained using scenarios.

Browsing the namespace

An *OPC Batch Client* uses the Browser Interface as defined in the *Data Access Specification* to browse the namespace. The namespace of an *OPC Batch Server* is always hierarchical. This is due to the kind of application involved. Browsing the namespace is carried out in the same way as namespace browsing of a *Data Access Server*. The principal procedure with browsing is described in section 2.3.2. Details of the OPC Batch namespace are explained further below. The client uses methods to move through the namespace, to request parts of the namespace, and to query for fully qualified ItemIds.

Applications of batch processing are always dynamical, i.e. devices, batches, or individual components can be added or removed. The client has to consider that namespace information can become invalid in the course of time and OPCItem objects can no longer be read or written. No new interfaces are defined to inform the client on such events, but the use of existing return values in the scope of the *OPC Batch Specification* is clearly defined. This is reflected in the following **Table 2.17**.

***Table 2.17:** Error definition in the OPC Batch Specification*

Error cause when browsing the Batch Server name space	Return value
Current position in the name space is invalid	Method calls return a general error
Identifier for modification of the position in the name space refers to an invalid position	Method calls return the error "invalid argument"
Browsing of the name space refers to identi fiers which no longer exist	Different return values of subsequent calls dependent on the modifications in the name space
Request of fully qualified ItemID with an iden tifier for a no longer existing leaf	Method calls return an OPC specific error code
Other methods (e.g. read and write) access to OPCItem objects which no longer exist	Method calls return an OPC specific error code

Definition of specific properties

The version 2.00 of the *Data Access Specification* defines the concept of properties at leaves in the namespace. This has also been extended to nodes in the namespace in version 2.05. A part of the overall value range of the PropertyId (100 – 4999) is reserved for specific OPC properties, which only need to be supported by items within a specific application range. Properties in this area are used within the *OPC Batch Specification* for mapping of relevant facts (e.g. information on resources) of the batch processing. For the specific nodes and leaves of the namespace, a large number of properties is defined which are situated in this value range of the PropertyID. *OPC Batch Specification* defines 57 properties. **Table 2.18** shows some examples.

Table 2.18: *Properties for items in the namespace of an OPC Batch Server*

PropertyID	Description
411	RelatedBatchIds; A string identifying a batch or list of batches. The physical model levels of Site, Area, and Process Cells are intended to return a list of BatchIDs related to their ModelLevels. Generally Unit, Equipment Module, and Control Modules return a single BatchID related to the batch using the Model Level, however when they are shared use resources they may return a list.
414	Location; A string representing a building or physical location where the item exists.
422	EquipmentState; A string representing the current state of the equipment.
437	ProductId; Product which the execution of this batch will produce.
446	ActualStartTime; Time when the batch, or other item, actually started.

After the client gets access to an item, it will query the list of existing properties and then read the values. It is also possible that OPCItem objects are created in OPCGroup objects for properties.

Besides these defined properties, vendor-specific properties (starting from PropertyID 5000) can also be available.

Access to parameters and results

Batches can contain parameters and results. *Parameters* are used to send values from a batch to a device. *Results* are used to send results arising from the execution of a batch in a device to the user (e.g. visualization). Parameters and results can exist at every level below the node *OPCBBatchModel* in the namespace. These items, in turn, can also have properties.

Some clients, for instance, are able to load all the parameter information into the devices when starting and then monitor the changes. Other clients can request specific parameters and then monitor specific results. There are two defined identifiers in the namespace – *OPCBParameters* and *OPCBResults* – which support this concept. The corresponding parameters and results are below these nodes. Parameters and results have a number of properties on their part (e.g. value).

Should it not be desired to browse every parameter/resultID, a so-called anonymous access can be used. The client requests the number of parameters/results and then addresses in the form of OPCBP[n].<propertyname> or OPCBR[n].<propertyname>.

"Propertyname" is used here for possible properties, e.g. value, access rights; *n* is the number of parameters or results.

Access to Summary BatchList Data

In cases of interaction with a batch system, one must check frequently to see which part of the application is executing. The client is interested in information on:

– the list of batches executed at a given moment

– the list of batches waiting to be executed

– the list of batches the execution of which is terminated

The client gets back a list of the BatchIds on demand. In the next step, the client will read the OPCItem object for the entry of the *OPCBatchIdList* in the namespace. The client is now able to request the status of the batch via the property "State". This is a very complicated way. It is more quickly by using the interface *IEnumOPCBatchSummary.* This interface enables the client to get all the information to batches with only one call.

Enumeration Concepts

OPC Batch Clients often want to record status information. To do this, enumerators should be used instead of text information. Enumerators are characterized by a identifier (e.g. OPCB_ENUM_PHYS) and a number (e.g. 0) describing the enumerator, and identifiers (e.g. OPCB_PHYS_ENTERPRISE, OPCB_PHYS_SITE) and numbers (e.g. 0, 1) describing individual values of the enumerator. The *OPC Batch Specification* reserves 99 possible enumerators. Some of the enumerators have already been defined. The client can ask the server which enumerators are supported.

Construction of fully qualified ItemIds by the OPC Batch Client

A large part of the namespace of an *OPC Batch Server* consists of the identifiers already known (OPCBBatchModel, ...) or can be derived from the contents of properties. Then, no browsing of the namespace is necessary to find out about the ItemIds. For constructing the fully qualified ItemId, the client has to know the server-specific delimiter. It can be queried by using the corresponding method.

Figure 2.23 shows the *OPCBatchServer* object. It has all the interfaces that have also been defined at the OPCServer object of the *Data Access Specification*. Note that the interface *IOPCBrowseServerAddressSpace* is not optional. Three additional interfaces have been added, providing access to functionality required for batch processing.

Methods for requesting the separators for fully qualified ItemIds and the generation of enumerators are available at the interface *IOPCBatchServer. IEnumOPCBatchSummary* provides the methods that enable access to the batch information in the enumerator created. *IOPCEnumerationSets* enables access to the enumerators explained above.

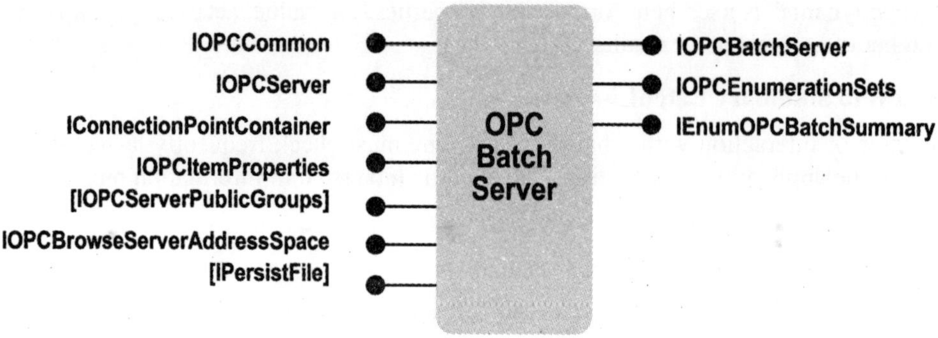

***Figure 2.23:** OPCBatchServer object in the Batch Server*

An *OPC Batch Server* also contains OPCGroup and OPCItem objects. These objects have the same semantics as apply with the *Data Access Server* and are therefore not illustrated. The same is true for the OPC object of the client.

Automation Interface

Automation objects of the *OPC Batch Specification*, on their part, are implemented in an automation wrapper. **Figure 2.24** shows the object model made available by the wrapper.

The object model is identical to the object model for the *Data Access Specification,* except that three objects have been added.

The following differences to the object model for the custom interfaces are evident:

- *OPCBatchSummaryCollection* object
- *OPCBatchSummary* objects

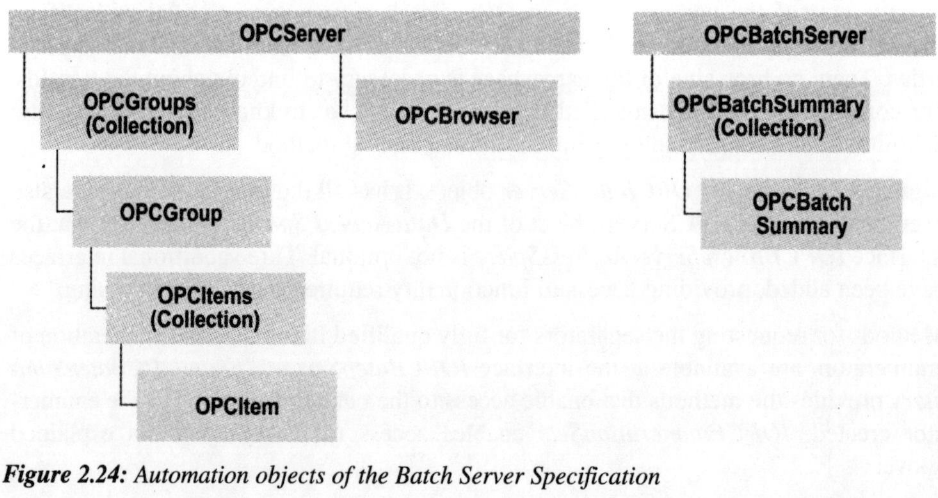

***Figure 2.24:** Automation objects of the Batch Server Specification*

2.12 OPC Security

2.12.1 Field of application and model

The descriptions given in this section only apply to DCOM security. Information on security in connection with Web services is provided in section 3.6.2.

The following information refers to version 1.0 of the *Security Specification*. The contents of this section and the next are closely linked to chapter 3.6. It is perhaps recommendable to return to this section after reading section 3.6.

All the specifications explained up to now define complete application programming interfaces between OPC client components and OPC server components. These components can run on different computers. Sometimes, security policies have to be set to restrict access to all the functionality and information in servers. Section 3.6 explains two ways to apply security specifications:

– Declarative security, by using the service program DCOMCNFG.

– Programmatic security, by using methods of the DCOM-Security-API.

Both approaches have advantages and disadvantages. The biggest disadvantage to applying DCOMCNFG consists in the large granularity of the security objects – all definitions are either made for all DCOM components on the computer or for a single component. For instance, it cannot be determined that specific methods can only be invoked by certain users. Here, the programmatic security is more flexible. There are no restrictions regarding the granularity of the security objects.

The *OPC Security Specification* concerns the use of the programmatic security. The disadvantage of the programmatic security is that interoperability problems may occur if it is used in a thoughtless and not agreed manner. These problems will be avoided by the definitions of the *OPC Security Specification*.

Security specifications are based on the definition of users having particular rights. As long as the client and the server are running in the same domain, no problems occur when defining these users – the domain users are used. However, this is not so easy if client and server are not running in the same domain. Users of one domain are not known in other domains. Introduction of specific users known only to OPC applications is a flexible solution.

This issue is also dealt with in the *Security Specification*. The *Security Specification* defines interfaces with methods, no new servers, or OPC objects. These interfaces can be integrated into servers based on existing and future specifications. The field of application of the *OPC Security Specification* can be compared with the field of application of the *OPC Common Definitions and Interfaces Specification*. Both define interfaces that can be implemented in all other servers. There is one difference, however – the interface *IOPCCommon* is a mandatory interface. The interfaces of the *OPC Security Specification* are optional.

The *Security Specification* is based on the Windows NT Security model shown in **Figure 2.25**. You must understand this model in order to understand the definitions of the specification described below and the explanations in section 3.6.

The model consists of the components described in **Table 2.19**.

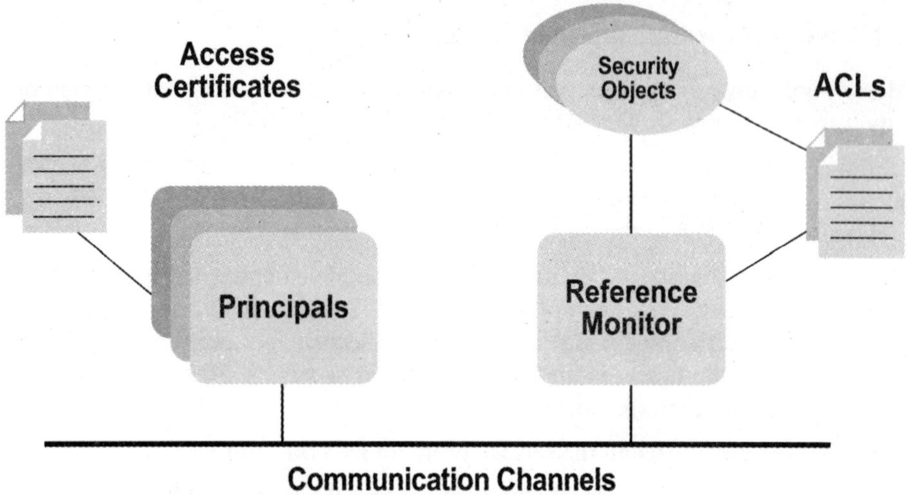

Figure 2.25: *Model of the NT Security*

Table 2.19: *Components of the NT Security model*

Principal	An active entity which has a need to access one or more security objects. Some examples of Principals are human users, computer systems, etc. In Windows NT, NT processes represent all Principals, which can be either interactive or service.
Access Certificate	A trusted credential associated with a Principal, which is used by a Reference Monitor as a basis for an access authorization decision. An example of an access certificate is the NT Access Token associated with every NT process.
Security Object	Any entity to which access is to be controlled. Some examples of security objects are files, directories, registry keys, key systems, etc.
Reference Monitor	An active entity which makes access authorization decisions for a set of security objects. A well-implemented security system ensures that all access requests for a security object are routed through its associated reference monitor, that is to say, no access to a security object can be granted except by the approval of its reference monitor. An example of a reference monitor is the NT reference moni tor for NTFS files and directories.
Access Control List (ACL)	A structure associated with every security object which specifies those Princi pals and groups of Principals allowed to access the secure object along with the types of access each is allowed. Examples of ACLs are the NT Access Control Lists that are associated with files, directories, DCOM servers, registry keys, etc.
Channel	A communication path between any two active entities within the system. A channel may be entirely contained within a single system, for example inter-process communications between two processes co-located on the same system. Alternatively, a channel may span computer systems using a commu nications network. A channel that is contained within a single computer system is typically considered to be secure. A channel that spans computer systems is typically considered to be insecure, unless specific actions have been taken to secure it. A channel may be secured to one of three levels: Authenticated A recipient can be confident of the identity of a message sender. Verifiable A recipient can be confident that the message has not been modified in transit. Private A sender can be confident that the message will be read only by the intended recipients.

The definitions of the *Security Specification* can be used in systems which:

- Require specific or flexible settings of security parameters (access to OPC Servers via the internet, modification of access rights during operation of the plant, implementation of role concepts)
- Which only partly support the model of Windows NT Security (Windows95/98, Windows CE and DCOM portation). These systems do not have all the functionality of the DCOM-Security-API, or they lack user credentials as defined by Windows NT.

The specification defines three security levels:

Disabled Security

- No security is supported. Every user can launch the OPC Server and has access to it. The OPC Server does not monitor access to manufacturer-specific security objects.

DCOM Security

- Only NT Security is supported. Start and access rights are limited to selected clients (users). The OPC Server does not monitor access to manufacturer-specific security objects. This is the default security level provided by DCOM.

OPC Security

- The OPC Server is the reference monitor which monitors access to manufacturer-specific security objects in the server. An OPC server can implement OPC-Security in addition to DCOM security or can only offer OPC-Security.

2.12.2 Specification object and use

In the standard case, an OPC server that has implemented security objects will carry out its access decision based on the NT Access Token (Credential). There can be cases, however, where NT Access Tokens are not available:

- The OPC server is running on a system that is not an NT operating system (e.g. Windows CE, Unix,).
- The client is running on a system that does not support generation of the NT Access Token.
- The concrete application requires the use of other credentials (access to OPC Servers from other domains via the Internet, role concepts).

For these cases, the specification provides the use of OPC credentials. The specification defines methods at the interface *IOPCSecurityNT* which are used in connection with the NT Access Token. Methods at the interface *IOPCSecurityPrivate* are used in the context of OPC credentials.

An explanation of the functionality of the methods is made by means of scenarios.

Use of Security Interfaces

An OPC Server can either implement the security interfaces or not. A client can query whether or not the interfaces are implemented and can then use them. In certain phases of OPC Server use (e.g. during commissioning), it can make sense to disable security functionality. Methods are available at both interfaces with which a client can query whether the corresponding security functionality can be used in the current configuration.

Access to security objects by using NT Access Tokens (Credentials)

During operation of the OPC servers, it is sometimes necessary to carry out certain modifications, such as the creation of OPCGroup and OPCItem objects in a *Data Access Server* for example. The user who has launched the client and has access to the server is possibly not entitled to do so. It could be possible, for instance, to temporarily add an OPCGroup object with defined status items for service purposes in a continuously running SCADA system. The service staff who check the plant daily do not have the required rights to do so. A technician will inform the OPC Client of his (additional) rights, his NT Access Token. The client will then inform the server of the modified credentials. The server will request the NT Access Token using the DCOM-Security-API functionality (see section 3.6) and use it for a further security control. After finishing this action, the technician will inform the client again of the use of the previously valid credentials. This information is then given to the server by the client. This flexible mechanism can be used in order to modify user rights at runtime.

Access to resources

For security reasons, the server process will not use the identity of the client process by which it is started, but will execute using a different user account. This means that the server is limited in the ways it can access other resources. The client is responsible for determining to what extent the server process can impersonate the client. This is called the impersonation level.

The server has to (partly) impersonate the client to carry out specific actions to which only the client is entitled, e.g. to access a file. The client is able to query the minimum impersonation level required for such actions from the server. The client can then modify it, if necessary. Section 3.6 gives a further description of the different impersonation levels.

Access to security objects using OPC credentials

OPC server access is to be possible from a different domain. In order to make access as flexible as possible, OPC credentials are used, not NT Access Tokens. NT Access Tokens must be known both in the domain of the server and in the domain of the client; when many different users are interested in launching the client, this is not practical. A solution is the use of OPC credentials that have been created for the specific purpose of accessing OPC servers. Several credentials can be defined, for which different rights are described in the Access Control Lists for the security objects in the server. These credentials have to be made known to the group of users who have access to the OPC server via the Internet.

The client launches the OPC server and accesses the interface that enables security decisions, which are based on OPC credentials. The client will transfer the OPC credential to the server. The difference between this scenario and the use of NT Access Tokens is that the credentials are now transmitted via the communication channel. It must be ensured that the information is not read or changed by unauthorized persons. To this end, specific actions have been taken to secure the communication channel. This is achieved by means of a method explained in section 3.6.

Interfaces

Figure 2.26 shows the defined interfaces. You can see that these are optional interfaces. "XXX" is used for "Event" (OPCEventServer), for "HDA" (OPCHDAServer), or it can be empty (OPCServer).

***Figure 2.26:** Custom Interfaces of the Security Specification*

Automation Interface

Work on this specification was not completed and therefore will not be covered here.

2.13 OPC Compliance Test

By Paul Hunkar and Jeff Harding, Compliance Working Group of the OPC Foundation

Open standards such as the OPC specifications are created with the aim of permitting interoperability between products from different manufacturers. In order to ensure a product actually complies with the specification it must undergo a series of tests. Users need to feel confidence that products they have purchased or intend to purchase from different manufacturers will work as expected when connected to one another. The user can gain this confidence by selecting certified products. At the time of this writing the OPC Foundation offers a self certification test that is run by server manufacturers and a series of Interoperability (IOP) workshops. The OPC Foundation is also in the process of creating a broader compliance test which would be carried out by an independent third party certification lab.

The non compliance of a product typically stems from either an incorrect interpretation of the specification or because of an incomplete or incorrect implementation of the specification. Compliance testing is designed to assist the manufacturer in finding and correcting non compliances before a user encounters them in the field. The Interoperability workshops allow a client vendor to test there implementation against existing vendor provided servers. The IOP workshops are also designed to assist the manufacturer, in particular Client Vendors, in the same manner as Compliance Testing does for server vendors.

Online Catalog

The OPC Foundation maintains an online catalog of vendor's products at its web site which can be found at (http://www.opcfoundation.org). The online catalog presents users and potential users of OPC products with descriptions of the various vendors' offerings. In addition to the descriptions of the products the compliance status of the products is also made available. Filters are provided to make it easy for user to focus on products that are certified. **Figure 2.27** shows an example screen shot of the catalog.

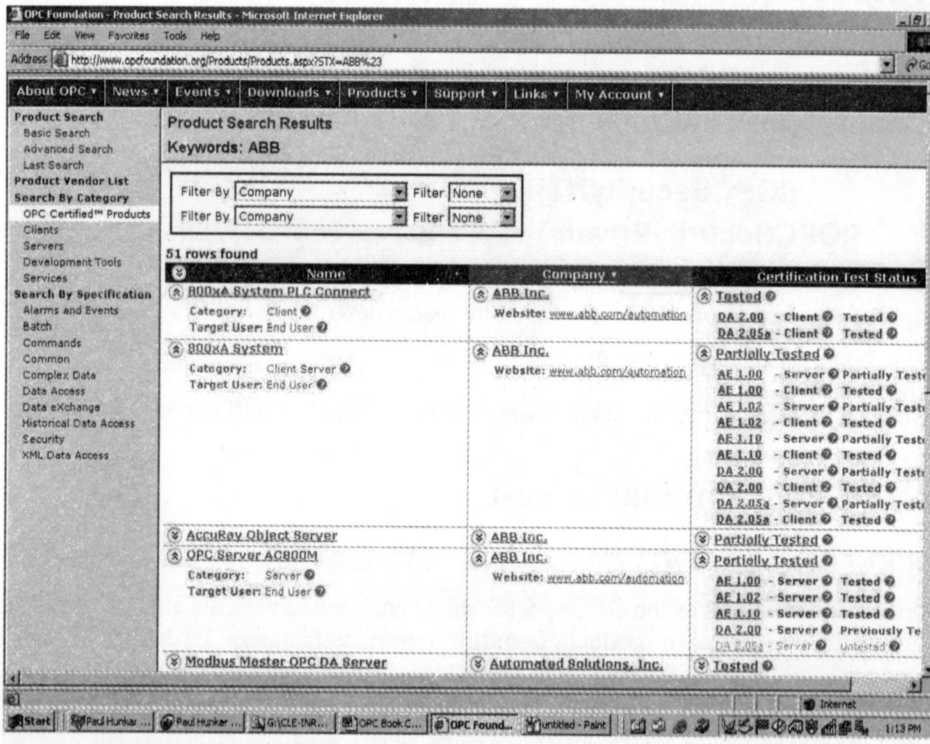

Figure 2.27: OPC Foundation Product Catalog

2.13.1 Self Certification

Self certification can be separated into two grouping, certification for server products and certification for client products. For Server vendors, the OPC Foundation has created and maintains a suite of self certification tests and makes them available to its members. The self tests are designed to verify a server is compliance with the various specifications that comprise the OPC standard.. For Client Vendors, the OPC Foundation conducts Interoperability (IOP) workshops three times a year (once in Europe, the US and Japan). These IOP workshops provide a client vendor the opportunity to check there implementation against standard server vendors in a semi structured manner.

Compliance Test Tools

The OPC foundation provides Compliance test tools. Manufactures of server products download and execute these tests in order to verify their products are compliant. The tests are provided in the form of an OPC client application which has been designed to automatically execute a series of verification test by interoperating with the server under test. The test client generates a test result that the vendor sends to the OPC Foundation for publication within the online catalog. To ensure the integrity of the test report it is encoded by the test client as it is created.

The self certification tests are designed to thoroughly test the operation of the interfaces that comprise the specification being certified. The tests fully verify that the interfaces of the server under test respond to client request in compliance with the specification. Despite the thoroughness of the tests the internal behavior of the server can only be verified to a certain extent.

The test cases are designed to verify the behavior of the server using both valid parameters when making calls as well as invalid parameters. It is important to not only verify that a server is compliant under normal conditions but also when a client is not as well behaved as it should be. For example tests are performed with a valid ServerHandle parameter as well as an invalid ServerHandle parameter. Other test cases attempt to write to OPC items that are known not to exist and verify that the correct error response is returned by the server. Repeatability is a key factor in the design of the test cases. Designing repeatability into the test cases ensures that any variation in results can be attributed to the server under test.

Test cases are designed to include verification of parameter range. Test cases will typically issue requests to the server using the low range, the high range, and some selection of parameter values within the range to ensure server compliance. Many tests can only be carried out after some preparation of the server's state. For example, it is only possible to read an OPC item after first creating an OPC Group and then adding the item to the group. After the test, these preparations must be reset in most cases in order to make the next test possible.

The Compliance Test tool supports the following OPC specifications at this time:

- OPC Alarm and Event 1.1
- OPC Data Access 2.05a
- OPC Data Access 3.0
- OPC Historical Data Access 1.2
- OPC XML Data Access 1.01

The User should first select the Specification that is to be tested by the tool. The available specification are listed under the "Extra" menu under "Options" (see **Figure 2.28**).

For each available specification the *Compliance Test Tool* supports the following test possibilities:

- Logical tests – These tests test whether the Proxy/Stub-DLLs are installed correctly.

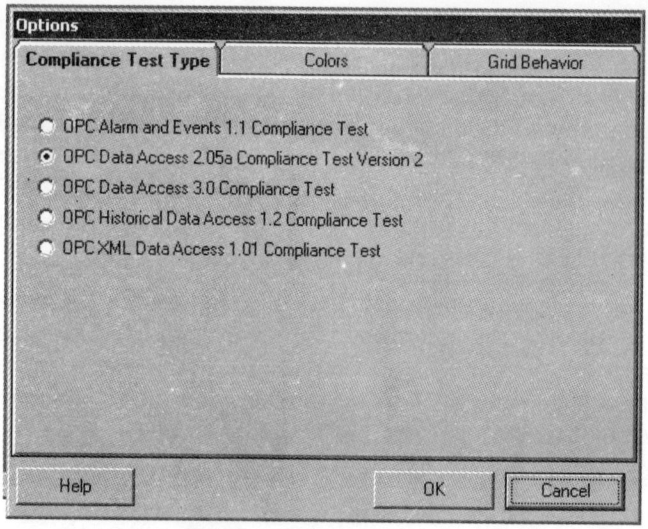

Figure 2.28: *Compliance Test Options*

- Interface tests – These tests test the methods that are part of the interfaces defined by the specification. Valid, invalid, and meaningless parameter values are used for all the methods.
- Stress Tests – For some specification stress tests are included.

Configuration of the *Compliance Test Tool* has to be carried out prior to starting the test run. In the following paragraphs the test procedure for each of the compliance tests is discussed.

The corresponding dialog page for the test of a *Data Access Server* can be seen in **Figure 2.29**.

First step is to select a *Data Access Server.* It can be located on the local computer or on a remote computer. The selected server is used both for interface and stress tests.

The "Test Items" tab enables the definition of identifiers in the namespace for which OPCItem objects are added during the test.

The next two tabs deal with the testing of the particular methods. Under the "OPC Server" tab, the items for which tests are carried out for the interface IOPCItemMgt are selected. Under the "OPCGroup" tab, the properties of the OPCGroup object that will be created and the monitoring times for carrying out the tests are defined.

The "Stress" tab allows users to set default values for the stress test. This tab allows users to determine how many OPCServer objects can be created in the component to be tested. In practice, this would mean that such a number of OPC clients communicate with the server. The same is also true for the number of OPCGroup objects per OPCServer object and the OPCItem objects per OPCGroup object. The maximum value for all the settings is 999.

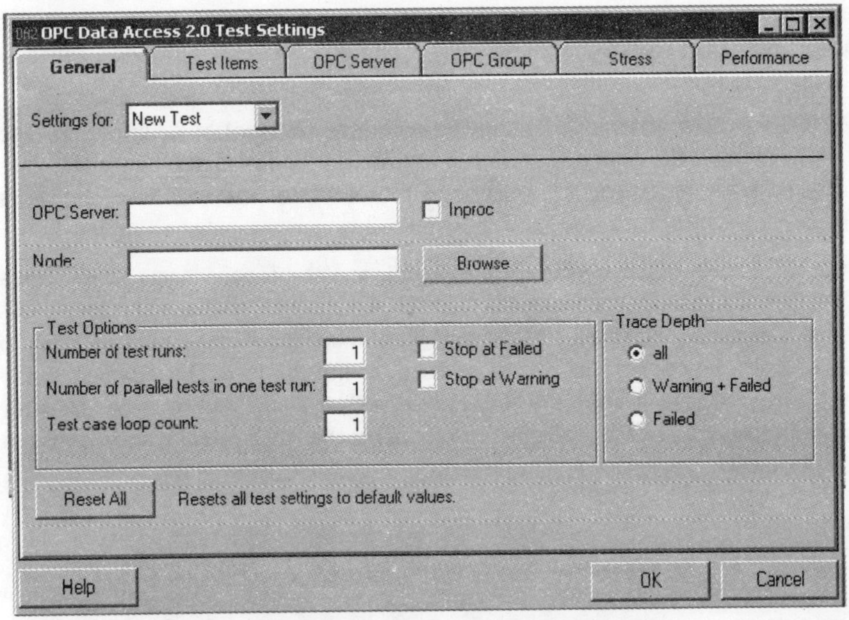

Figure 2.29: Configuration dialog page for the DA Compliance Test

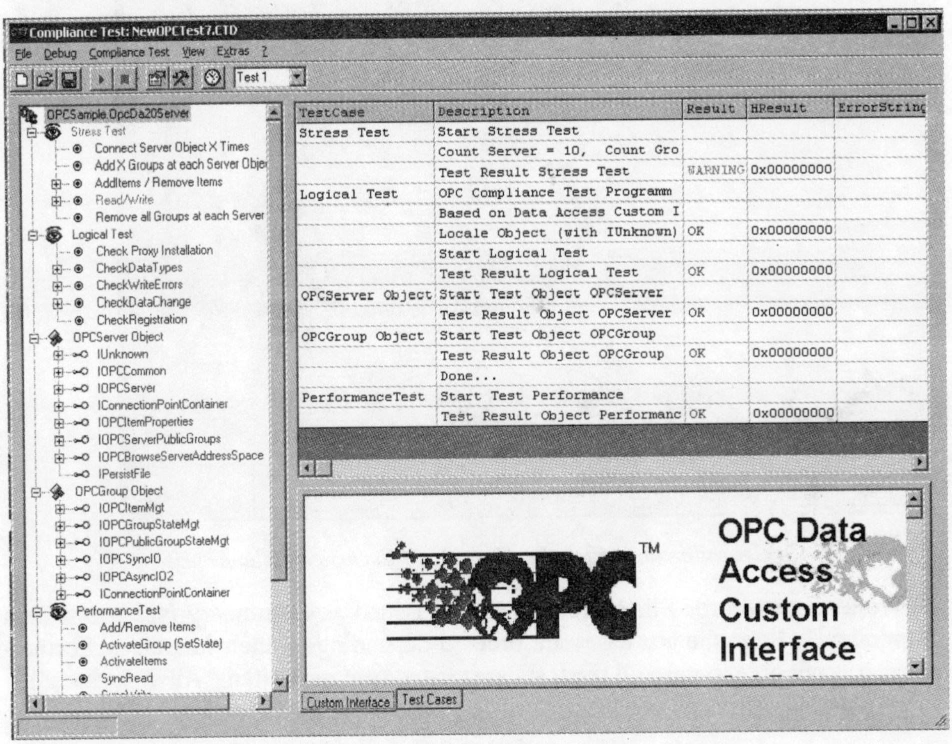

Figure 2.30: Summary of the results of the DA test cases

The results are recorded in a file and can be issued as a summary (see menu item "Summary"). Here, the test cases are ordered depending on their interfaces. Furthermore, it is possible to have all the test cases listed (see menu item "All Test cases")

Figure 2.30 shows the information displayed when the test tool has completed. The various test results are illustrated by different colors. In the summary, every test case can be chosen. A short indication is given of the object of the test case, the input values, the expected return values, and the result.

The **Figure 2.31** shows the configuration dialog a test of an *Alarms and Events Server.* During configuration of the A&E test, the A&E server that is to be tested has to be defined. Once a server is defined the user should use the Areas and Source tab to select Alarm and Event areas that are to be used during the test. The AE Server tab should be used to define a source node and condition event that will be used for condition event testing. The Compliance Test Tool for the A&E specification does not provide a stress test.

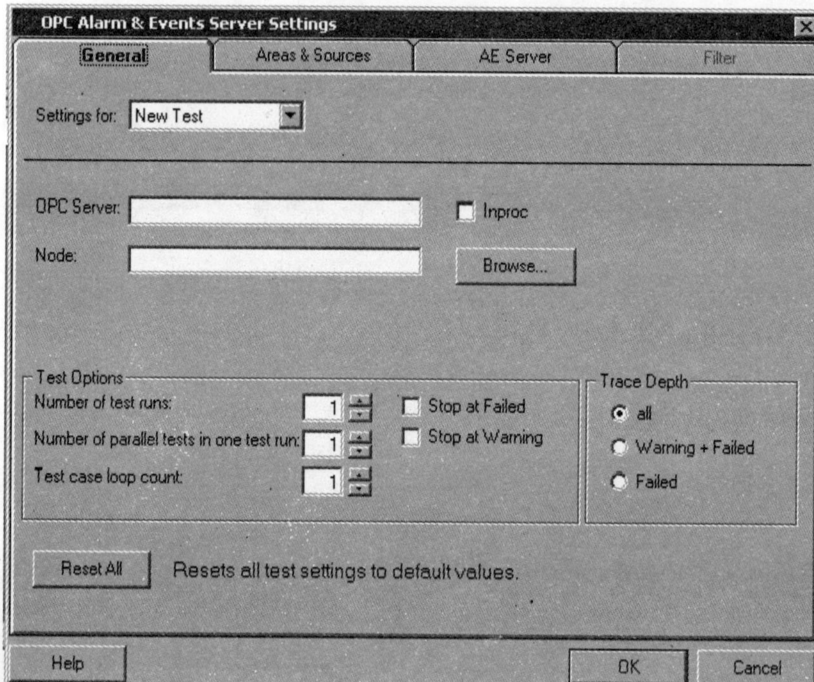

Figure 2.31: Configuration dialog for the Alarms and Events compliance test

The results are recorded in a file and can be issued as a summary (see menu item "Summary"). Here, the test cases are ordered depending on their interfaces. Furthermore, it is possible to have all the test cases listed (see menu item "All Test cases")

Figure 2.32 shows the information displayed when the test tool has completed. The various test results are illustrated by different colors. In the summary, every test case can be chosen. An additional screen is generated during the run that shows the events that were generated.

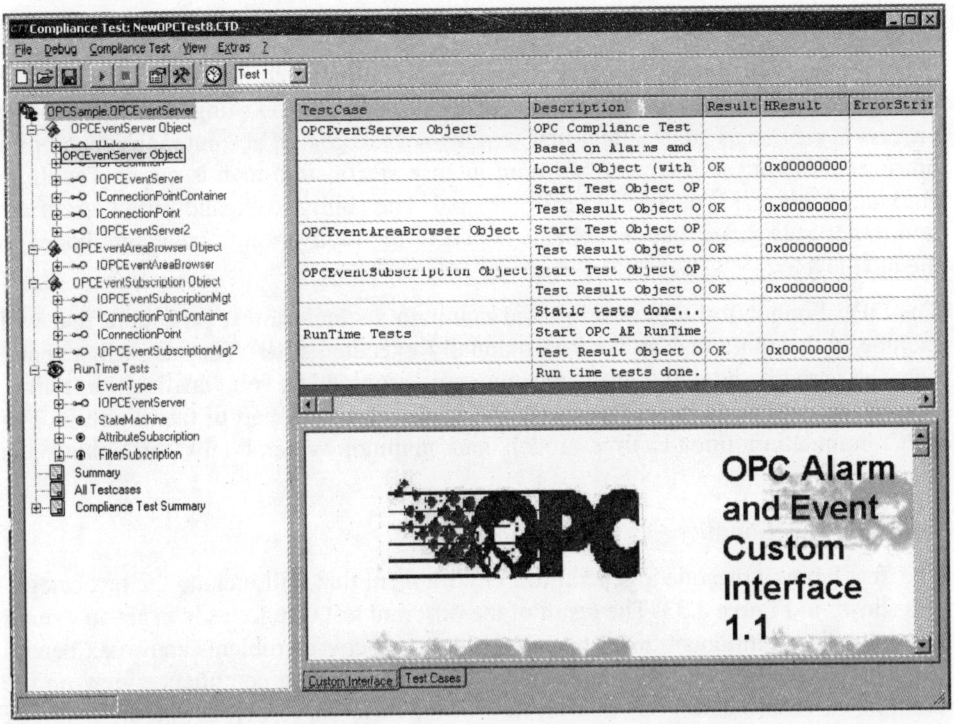

Figure 2.32: Compliance Test Tool for a test of an Alarms and Events Server

The tests of a *Historical Data Access Server* and a *XML-DA Server* with the Compliance Test tool are similar to the tests of a *Data Access Server* or *Alarms and Events Server.*

2.13.2 Interoperability Workshop

The OPC Foundation conducts Interoperability (IOP) workshops three times a year (Once in Europe, the US and Japan) These IOP workshops provide a client vendor the opportunity to check their implementation against standard server vendors in a semi structured manner. Attendance at the interoperability session is strictly voluntary. The interoperability test process requires that OPC vendors run a series of standard tests for each combination of an OPC Client and an OPC Server product.

This is the only method for client to be tested. It does have some limitations, in that the subset of interface under test is limited and the testing of the interfaces is limited. It is not a full coverage test. The environment changes every session, since different vendors participate. The testing is very informal self testing by the vendors. There are no checks to verify if test cases are actually executed. Also since vendor provided servers have usually passed compliance tests prior to an IOP, they are well behaved and clients are not tested for ability to handle even simple errors.

2.13.3 Third Party Certification

At the time of this writing the OPC Foundation's compliance working group is in the process of defining an additional certification level. The existing self certification process is viewed as an excellent first step toward the goal of absolute interoperability however it is not the final step. A more intense verification than is possible with an automated self test is needed to fulfill the goal. The ability to expand formal certification to not only server but also clients is part of the motivation to take compliance to the next level.

The OPC Foundation will develop and maintain a standard test procedure that will include both automated as well as manually executed tests. The test will expand both the breadth and depth of the test coverage provided by self certification testing. Test lab accreditation procedures will also be developed as part of the initiative. The OPC Foundation intends to establish and maintain several labs throughout the world.

Test Lab Functionality

The Test lab will provide a general test environment that will include the test categories shown in **Figure 2.33**. The result of the different test categories is to get an overall picture of the component under test as well as to discover problems and weak behavior. The goal is to raise the quality of the testing, beyond a compliance view on the component. **Table 2.20** gives an overview of the different test types and the coverage with the OPC Certification Modules.

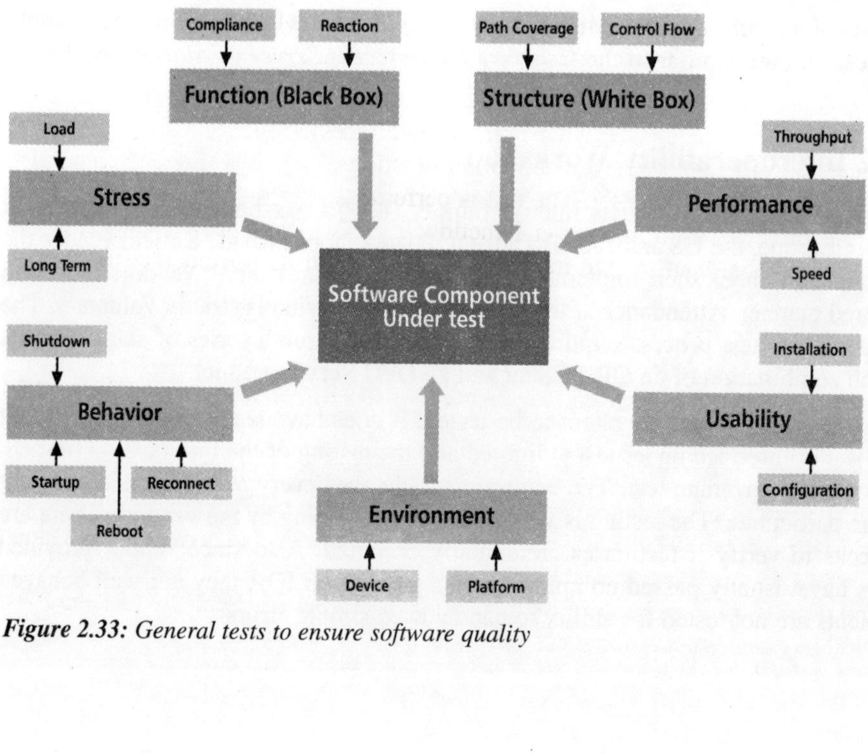

Figure 2.33: General tests to ensure software quality

Table 2.20: *Overview test coverage OPC Server*

Assignment	**Compliance**	**Interoperability**	**Test Lab**
Black Box Testing	Full	Partial	Full
Performance Testing	Partial	–	Almost Full
Load Testing	Partial	Partial	Full
White Box Testing	–	–	optional
Behavior	–	Little	Full
Environment	–	Partial	Almost Full
Usability	–	–	Partial

The seven main categories of tests to be performed will cover all essential aspects of an OPC product. Whether or not these categories are in the scope of certification is still being determined at the time of this writing. The Independent Test Lab will offer all of the categories, but some such "Structure" may only be offered as a service and not considered as part of the Certification.

Test Lab Functionality

When a product schedules time in a certification test lab, a fixed environment will be provided. The Test Lab will have a consistent environment and procedures in place for both Client and Server testing. The following figures illustrate the possible environment for both a typical server test and a typical client test.

Server

The tester performs the test sequence regarding the list of test cases. **Figure 2.34** shows the environment for an OPC Server product.

The OPC Server vendor provides the test candidate and a list of the server's functionality (What specifications and what optional interfaces). Based on this list the test cases which have dependencies to supported functionality will be executed e.g. OPC methods provided by the server. The test is performed with the list of test cases that can be executed based on the Server Functionality List. The disabled test cases (not supported) and the result of the executed test cases will be documented in the test report.

Client

The tester performs the test sequence regarding the list of test cases. **Figure 2.35** shows the environment for an OPC Client product.

The OPC Client vendor provides the test candidate and a check list of the client functionality. This check list is based on the test cases which have dependencies to supported functionality e.g. OPC methods used by the client. The test is performed with the list of test cases that can be executed based on the client functionality check list. The disabled test cases (not used functions) and the result of the executed test cases will be documented in the test report.

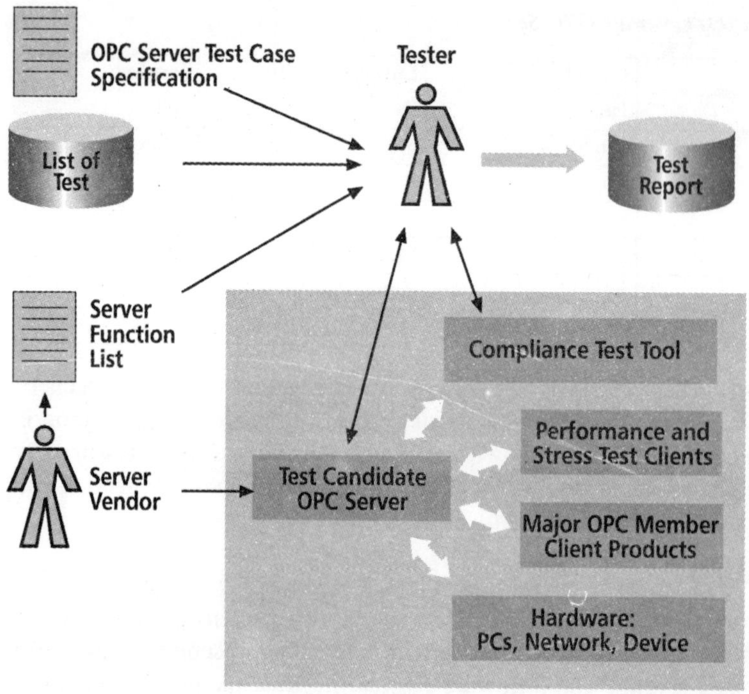

Figure 2.34: *OPC Server test environment*

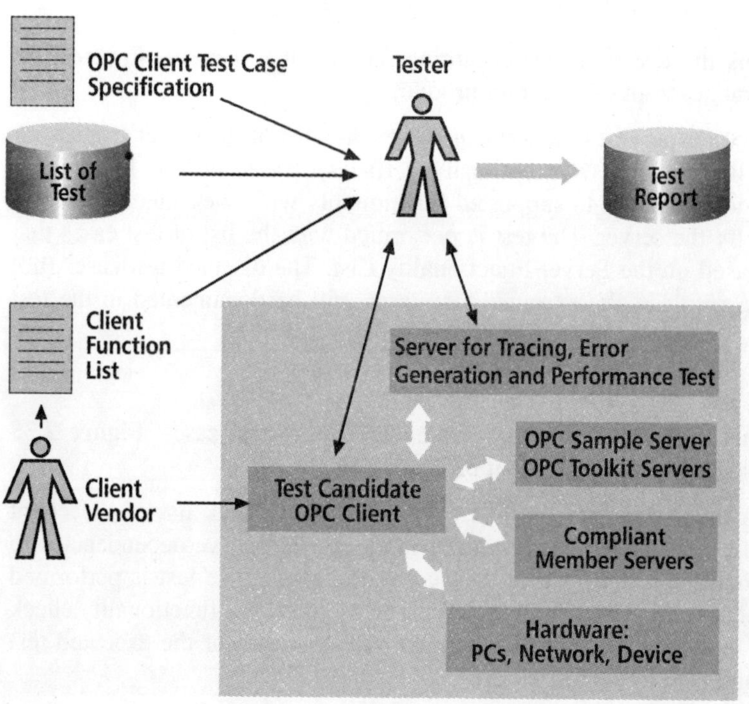

Figure 2.35: *OPC Client test environment*

OPC Certification

Testing will result in certificates that products will be able to display. The following steps illustrate the testing a typical product should follow. After passing the vendor's system test the OPC product should be self tested using the CTT (Compliance Test Tool for servers only). Additionally the product should be taken to an IOP (Interoperability Workshop for servers and clients). An OPC product that passed the IOP and the CTT will be marked as "self tested" and the results will be published on the OPC Foundation's web site. After being "self tested" the OPC product should be taken to an accredited OPC Test Laboratory. When passing the tests of the independent Lab the product will be marked as "OPC Certified". The certificate will be published on the OPC Foundation's web site.

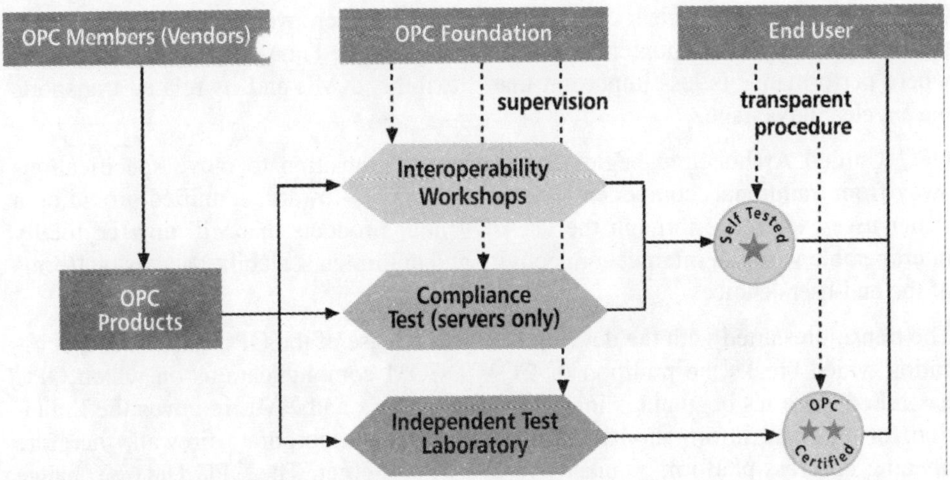

Figure 2.36: OPC Certification Testing

2.14 OPC Unified Architecture

By Tom Burke, president of the OPC Foundation

Members of the OPC Foundation working on the next generation of OPC Alarm & Events were bantering about in 2003 about Alarm & Events functionality and how to move forward the OPC Alarm & Events specification to web services when the first visions of the OPC Unified Architecture was really crafted out.

The process of looking at what was common across the OPC domain specification (data access, Alarm & Events, historical data access) was pushed into the OPC common specification as the "base" for all OPC Domain specifications.

The OPC specifications have been built primarily on the COM framework/Object Model which has the following advantages

– It worked
– It had huge market presence

– It was reasonably stable and well specified

– When used properly it provided very high performance as demonstrated by various benchmarks

The down side of COM/DCOM are the limitations of client and server applications being required to be on the intranet (inside the corporate firewall), and the lack of interoperability to platforms not COM/DCOM based including some embedded platforms and enterprise level server machines.

The OPC Foundation started leveraging and moving toward standards that supported XML and web services, with the first specification started being the OPC XML Data Access specification, followed by the OPC Data eXchange specification. In the short term the OPC Foundation still believes that for low level, high performance applications, COM remains the clear choice. However for higher level communications over the Internet and for communications with loosely coupled non-Microsoft applications where performance is less important than flexibility XML and its related transports have a clear advantage.

OPC Unified Architecture begins the process of transition to move specifications away from traditional 'component'-based connectivity towards a unified providing a richer user experience through the use of vendor products that will now be totally interoperable with full internet connectivity and unlimited scalability across platforms of the end-users choice.

The transition started with the development and release of the OPC XML DA specification which breaks the tradition of COM/DCOM communications on which OPC has relied since it's beginnings in 1995. Web services and XML removes the limitation for data and information to be isolated behind a corporate firewall, therefore opening up cross-platform connectivity over the Internet. The OPC Data eXchange specification, announced earlier in 2003, leveraged the use of web services in addition to supporting the traditional 'component'-based connectivity, thereby providing a choice of both strategies for cross-fieldbus interoperability via Ethernet. OPC data eXchange promotes interoperability between devices on the industrial Ethernet, and leverages the use of some of the facilities of the unified architecture using XML as the key technology for successful adoption and deployment.

OPC has traditionally provided interoperability between applications only inside the corporate intranet, allowing one application to exchange data with another application using Microsoft COM/DCOM framework. The server application identifies what 'components·that it wishes to expose to client applications and then the client subscribes to the components that it wants to be notified on exception. To move the exception-based data outside of the corporate intranet custom applications are required to become the bridges and gateways. These custom applications must expose all the components from the factory floor to the higher-level applications and data is exchanged; in reality, the higher-level applications are looking for information to be constructed from the data from the factory floor.

With the unified architecture, reliability, redundancy, security and information throughput is increased, as well as ensuring easier development and integration. It

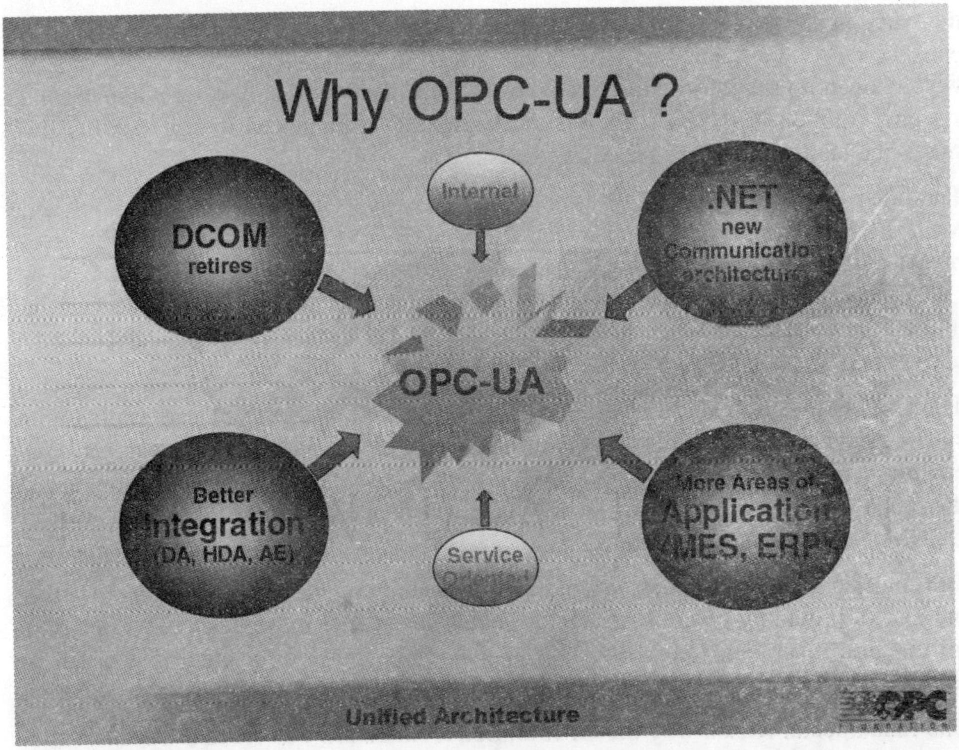

Figure 2.37: New requirements for the use of OPC and changing basic technologies lead to the OPC Unified Architecture initiative.

also means that raw data – which few people outside the plant floor are interested in – can be processed into meaningful information that meets the needs of managers, maintenance staff, accountants, etc. Asset Management and Production Control are just two areas that will benefit.

The OPC 'component'-based approach will remain valid indefinitely as the more than 7500 OPC products and many millions of installed systems remain in use. However, the OPC Foundation expects suppliers to quickly start using the services of the unified architecture for new product launches. Existing 'component'-based OPC products will be able to plug and play into the new Web services part of the unified architecture, allowing existing products that previously were limited to the corporate intranet to transform their data into information outside those boundaries, as well as be interoperable on a multitude of platforms and address the critical issue of scalability.

The transition and need has been coming for some time, as OPC responds to break-through technologies, particularly Microsoft's .NET environment. "the OPC Foundation mission is to deliver the very latest benefits to members, who basically are the suppliers of OPC functionality to the world's automation users. The trend towards a unified architectural environment has been clear for some time, as end-users have been demanding interoperability between vendor applications in support of building complex systems that span multiple platforms and the internet. What OPC brings to the party is an open methodology for delivering the benefits to everyone. This will enhance existing systems that have been built from 'components' by the end user as well as open up the information architecture from vendors that have provided a total systems solution in the past. This is the beginning of enhanced interoperability between the factory floor and the enterprise."

The other area being opened up is the middleware ground between plant floor and MES, ERP environments, which is characterized by UNIX platforms and a high level of consultancy. "This environment can already use OPC XML DA to connect with OPC systems". A standardized set of services that will be the key deliverables of the unified architecture will ensure maximum benefit for implementers, integrators and end users of these systems too"

The OPC Foundation could have taken the fragmented approach of just migrating a subset of the existing OPC DCOM Alarm & Events specification to the web services environment and creating an OPC XML Alarm & Events specification following in the footsteps of how OPC XML Data Access was created from the OPC Data Access DCOM specification.

OPC unified architecture is clearly in the direction of embracing Web services for the purpose of providing secure transaction reliable services to move data (converted into information) from the factory floor device and applications to the enterprise applications.

The fundamental reason why the OPC unified architecture was created was basically doing to address the need to migrate away from DCOM based architecture which essentially was the technology of the mid-1990s. The migration from the dcom based architecture was also necessary to address the need for integration of complex systems that span the corporate firewall boundaries. OPC leveraged the XML architecture in late 1990s to demonstrate the use of web services for the fundamental data access as the first migration from the dcom based architecture. OPC began to leverage Microsoft .NET framework and developed a .NET API for data access, historical data access, alarms and events. The .NET architecture was considered to be the first communication architecture providing the integration between Web services and DCOM. OPC also learned move in the direction of expanding the application boundary to move into the MES and ERP domains for integration between the automation applications and the enterprise applications. Part of the unified architecture is also driven by the need to tightly integrate all the OPC specifications have been developed and adopted over the last ten years.

The key features of the OPC unified architecture are:

Broad Application Scope:

With the COM specifications OPC was mostly confined to Microsoft Windows and hence the desktop. Since OPC-UA is Web Services based, it will be easier to run UA in devices (PLCs, controllers, DCS ...). On the other extreme high end enterprise systems running UNIX and other operating systems will also be able to host OPC-UA applications.

Open Communications:

OPC-UA utilizes the very latest Web Services technology including some of the latest WS-* standards. This gives UA great cross platform capability and lots of high level tool support.

Integrated Address Space and Object Model:

All of the functionality of the existing specifications can be implemented using the base services defined in UA. The variations in the functionality is confined to differences in the data, not differences in the interface methods.

Rich Information Model:

Another major departure from existing OPC implementations is how rich object based data from underlying systems is exposed. Currently individual simple data elements are exposed. With OPC-UA the native complex object will be exposed "as-is" without breaking them down into simple items.

The OPC unified architecture leverages the Web services standards inclusive of the major Web services interoperability (called WS-I) the address all the areas of functionality particular to the Web services to maintain for interoperability.

For example the WS-I specification called WS-Policy will allow OPC-UA clients and servers to negotiate among supported protocols and encodings at runtime. This will insure the use of optimal communications when possible while preserving the highest levels of interoperability.

For example the WS-I specification called WS-Eventing will allow to support true callbacks as used in the COM interfaces and will also support "polled refresh" introduced with XML-DA.

The OPC unified architecture will also focus at the needs of performance and scalability as part of the base services include the use of binary encodings in the Web service model allowing OPC unified architecture applications the same level of throughput of performance if not greater than the application counterparts that were based on the dcom architecture.

The design of the OPC unified architecture is all about building a solid foundation that supports the functionality of the existing OPC specifications (from both a functional and implementations standpoint). The OPC unified architecture has defined a base of services that represents the common functionality that is necessary for the existing the domains specifications as well as the future information model collaboration opportunities where OPC provides the services necessary to be the "HOW" for moving "WHAT". The OPC foundation is not in the business of having to know all the possible data/information in order to provide the services necessary for things like data acquisition systems, rather the focus is to provide generic interoperability services that I was other applications to discover the information and access the information using the normal OPC you put architecture base services for things like data acquisition and event notification.

The picture below begins to describe the picture how the OPC unified architecture base services are the foundation and than the OPC data access, alarm and event, historical data access and command functionality are built on top of those base services providing additional services as required for that particular domain.

OPC wants to be the foundation for reliability and interoperability in a secure operating environment providing set of services facilitating end users the ability to buy components from many different vendor's can be able to deploy those components into a total systems solution without requiring any level of integration services are custom solutions being developed to tie the generic components/applications together.

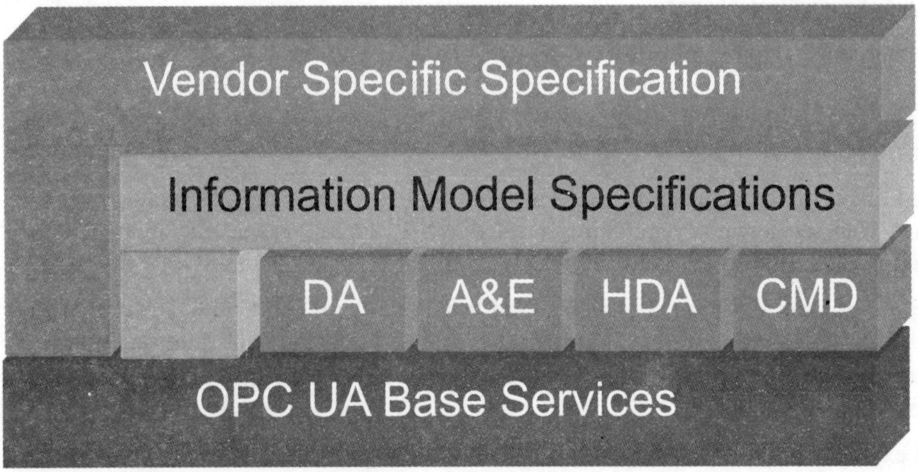

Figure 2.38: *The OPC UA services can be used by OPC applications and by applications from different standardization organizations.*

2.15 Example for using the specifications

This section describes the use of most of the specifications explained in the previous sections.

The section makes use of an hypothetical application, which is used for the control and monitoring of a building. The building has several floors in which rooms and halls are located. Process variables such as temperature, pressure, humidity, and flow are measured and, in some cases, controlled. To do this, measuring units and control units of various manufacturers are used. Besides the control task, this application is also used for safety monitoring – fire detectors and intrusion detectors for example. Variables are constantly logged and can be evaluated at a later date. There is a control room in the building where all the data are collected. Furthermore, it is possible to monitor parts of the application in individual parts of the building. Specific parts of the application are safety-relevant and can only be influenced by authorized staff. In principle, it is also possible to solve specific monitoring tasks via the Internet.

Thus, different clients and servers will be used which are implemented as defined in the *Data Access, Alarms and Events, Historical Data Access, Security* and *OPC XML-DA Specifications.*

Data Access Servers will enable read and write access to the measuring units and control units in interaction with the clients. The measuring units come from different manufacturers and the OPC servers are also made available by different manufacturers. These OPC servers have all to be adapted to the specific application. This concerns both parameter settings for communication between the PC and the devices and construction of the namespace in a *Data Access Server.*

Alarms and Events Servers are used for monitoring events. Events could be the exceeding of temperature or humidity limit values. Events should also be triggered if temperature setpoints are changed. A device failure is also reported as an event. Of course, safety aspects have to be mapped to events, e.g. open doors and windows, movements in rooms supposed to be empty, and so forth. The *Alarms and Events Servers* also have to be adapted to concrete situations. This concerns construction of the eventarea and the filter space.

The values for temperature, humidity, etc. have to be stored in a database and can be made available at a later date by a *Historical Data Access Server* to a *Historical Data Access Client* for analysis.

Some security aspects (permission for launching and accessing OPC Servers) can be influenced using the service program DCOMCNFG. Servers and clients will implement functionality of the *Security Specification* as only authorized staff should be allowed to change setpoints. Access via the Internet should be possible.

In a web server, a superserver will be running, which represents the interface between web applications and the OPC servers. Here, too, security aspects will be considered.

After the principal structure of the complete application has been presented, explanation of each aspects follows.

The different process variables are sampled by various devices at different locations in the building. This means that different OPC servers have to be used. There are three possibilities of deployment:

- *Data Access Servers* read variables from devices. A *Data Access Client* integrated within an *Alarms and Events Server* interacts with *Data Access Servers* to receive variables of interest. This can lead to an increased data volume but it can also be an advantage if only a small number of events are to be monitored.
- *Data Access Servers* and *Alarms and Events Servers* record variables from devices in parallel. This, however, may not be possible for some communication systems. This solution leads to additional communication traffic between the field devices and the PC.
- *Data Access Servers* and *Alarms and Events Servers* are located in one component. Data from the devices are only acquired once with this solution. If the volume of data is very large, server performance could be affected.

After decisions have been made regarding products and their deployment, the servers have to be installed and put into operation. During installation, attention has to be paid to the fact that the commonly used OPC components are installed according to the definitions in the *OPC Common Definitions and Interfaces Specification*. This is true for the correct registration and installation of the files.

The configuration of the OPC servers is the next step. The servers have to be adapted to the specific requirements of this application (e.g. communication parameters). Furthermore, OPC-specific structures (namespace, eventarea, etc.) have to be created. Most of the servers support these tasks very comfortably. For the servers, the corresponding settings have to be carried out for access and launch rights by means of DCOMCNFG.

After server configuration has been finished, the application on the client side has to be defined. The use of an OPC test client is strongly recommended to facilitate commissioning.

The test client makes it possible to check whether access to servers is possible and whether, for example, OPC objects can be added. After successfully finishing this test and solving the existing problems, the visualization application can be created by means of the client programs.

As an integral part of the visualization application, it has to be determined which data should be stored in the historical database. Implementation of the database itself depends on the concrete instance of the *Historical Data Access Server*. In addition, the namespace of the *Historical Data Access Server* has to be constructed. Furthermore, the aggregates to be supported must be determined.

Then, a HDA client is able to access the server. Similar to the access to *Data Access Servers* and *Alarms and Events Servers*, the client has to be configured and security aspects have to be determined.

On the same level, the XML server is to be put into operation to enable access of external clients.

It should be possible to write and read specific values not only from the control room but also locally. To support this, however, corresponding authorizations are necessary. Thus, security policies have to be set up that assign specific rights to various users. Configuration of the corresponding settings has to be made in the Access Control Lists of the servers.

Preparation of the OPC application is now finished.

3 Implementation

3.1 Introduction

Having dealt with the specifications in the previous chapter, this chapter discusses the implementation aspects.

As explained earlier, most of the OPC specifications released till today (October 2005) are based on DCOM, the Microsoft technology for the implementation of distributed, object-oriented applications. The chapter starts with an introduction to some aspects of this technology. The aim of this section is to explain the basic terms and to illustrate the correlations between them. For those who would like or who need in-depth information on DCOM programming, many books covering this subject in detail are available on the market. Many future OPC specifications will be based on Web services. Section 3.4 provides an introduction to this technology. As a number of specifications uses XML to describe data, this technology will also be covered by section 3.4.

OPC client and server components can be implemented in different ways and in a wide variety of development environments. Sections 3.2 and 3.5 will deal with creating clients and servers with the help of Microsoft Visual C++, toolkits, Visual Basic, and ActiveX Controls. You should have a working knowledge of programming in C++ and Visual Basic in order to understand the examples that will come up.

With the introduction of .NET technology in July 2000, Microsoft took another step in the development and use of platform-independent, component-based distributed systems. Section 3.3 provides a short overview of .NET and explains the importance of this technology for the further use of OPC in automation.

OPC client and server products can be used in various configurations. If client and server do not run on the same PC, it is essential that the user thoroughly considers the security aspects, regardless of whether the actual implementation of security requirements or only proper functioning is intended. The implementation of security aspects differs for Web services and DCOM. Section 3.6 provides an overview of the declarative security in DCOM and the use of security functions offered by the DCOM-API. This section also discusses the security aspects of Web services.

The last section covers the implementation and use of OPC on operating systems other than Windows 9X/ME/NT/2000/XP. Section 3.7 provides an overview of the points that need to be observed.

3.2 Distributed Component Object Model (DCOM)

3.2.1 Basic knowledge

History

Microsoft's DCOM is the result of two paths of development which are shown in **Figure 3.1**.

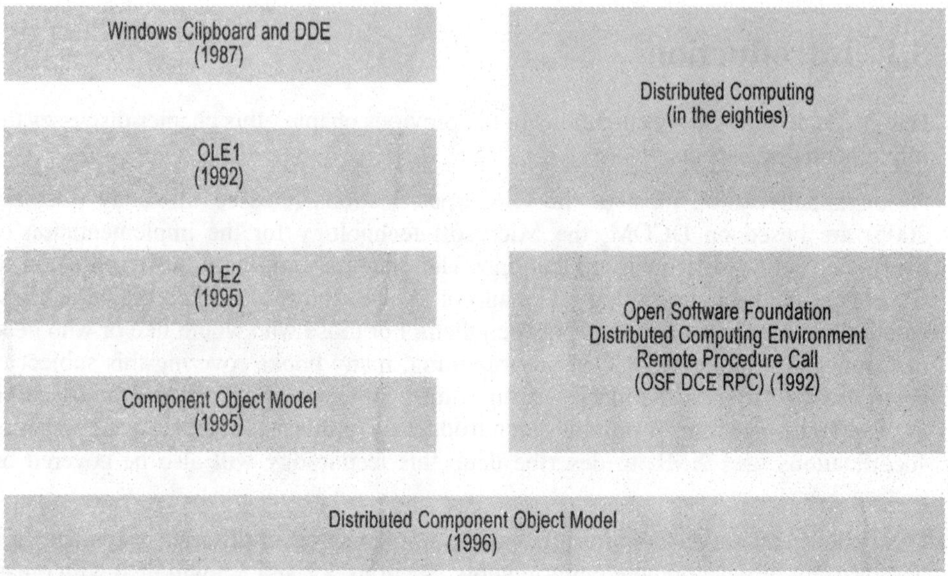

Figure 3.1: DCOM development history

The Windows Clipboard was the first solution in the Microsoft world with which information could be transferred from one application to another. It was based on the simple mechanism of "copy & paste" and was easy to use and understand. Dynamic Data Exchange (DDE) was another approach to exchanging data between applications. This technology, however, proved more difficult to use.

The clipboard mechanism allowed the creation of compound documents. This means that you could copy an Excel table into a Word document, for example. The problem was that this connection was not dynamic, not intelligent. Changes were not transferred automatically. To meet this requirement, Microsoft developed OLE1. OLE stands for Object Linking and Embedding, a technology that allows "living" connections between objects. Now it was also possible to have changes in the Excel table automatically transferred to the Word document (linking) and to edit them directly from there (embedding). DDE served as the communication protocol. The development of OLE2 took the evolution one step further – to component-based software. The technology was no longer simply intended for the interaction between text processing programs, but could be applied to any kind of interaction between a variety of components, since the model was now available to everyone. This technology has been known as the Component Object Model (COM) since 1995 and represents

Microsoft's basic technology for the implementation of (distributed) object-oriented applications.

Available since 1996, DCOM (Distributed Component Object Model) allows interaction with components on other PCs as well.

DCOM, however, also has another root; that is, in the development of Distributed Computing.

In the 1980s, companies joined up in the Open Software Foundation (OSF) with the aim of making distributed computing possible in heterogeneous environments. A heterogeneous environment means that the applications can be created in different programming languages, that they can run on different PC architectures, and that they can interact over various networks. This initiative gave rise to several specifications in the area of distributed computing, which have been summarized under the term Distributed Computing Environment (DCE). OSF solutions are still used in products today.

One result of the OSF's specification work was the definition of communication mechanisms between applications on different PCs. In this context, it is particularly important to determine how the different applications will understand each other and how each can call functions in the other. This also includes the following aspects:

- Transformation of parameters from a process-dependent representation into a representation that is independent of a specific process, operating system, and network. This is also referred to as marshalling/demarshalling.
- Addressing of methods in applications.

Two components – proxy and stub – are responsible for marshalling/demarshalling. A client does not interact directly with a server but calls the method in the proxy. Here, the marshalling takes place. The call is then passed on to the stub via the runtime environment. In the stub, the demarshalling is carried out and the method call is passed on to implementation in the server. When a callback method is accessed, processing is performed in the opposite direction. The manner in which individual data types (such as 2-byte integer, 4-byte integer) are converted is fixed. These data types are always combined differently, however, (e.g. arrays, structures) depending on the specific application. Proxy and stub components supporting this specific kind of marshalling/demarshalling therefore always need to be newly created.

The efforts to define communication mechanisms resulted in the OSF DCE RPC (Remote Procedure Call). DCE is not object-oriented. DCOM, however, is object-oriented. Microsoft has added the corresponding aspects to the OSF RPC, which has led to the designation ORPC (Object RPC).

Basic DCOM features

Interface

A DCOM object makes its functionality and data available via methods at interfaces. An interface is a strongly typed convention between a software component and a client. The interface provides a relatively small amount of semantically related operations. *Strongly typed* in this context means that there is a precise convention between client and server on how methods are to be called. A distinction is made here between standard and custom interfaces. Standard interfaces have already been defined by Microsoft as part of COM. There are, for example, standard interfaces with methods for memory management (IMalloc) and interfaces for the persistent storage of information (IPersistFile), among others.

The interface IUnknown, which has also been defined in COM, has a special meaning. Every standard or custom interface is derived from IUnknown and therefore offers the three methods defined for this interface in addition to other methods.

- QueryInterface – is used for querying whether a specific interface is supported by the DCOM object.
- AddRef and Release – are used for managing the life cycle of the DCOM object.

This all sounds very abstract; how about the concrete implementation?

In C++, an interface is very easy to define:

```
class IThisInterface : public IUnknown
{
    public:
    virtual HRESULT __stdcall DoSomeThing(int input, int* output) = 0;
};
```

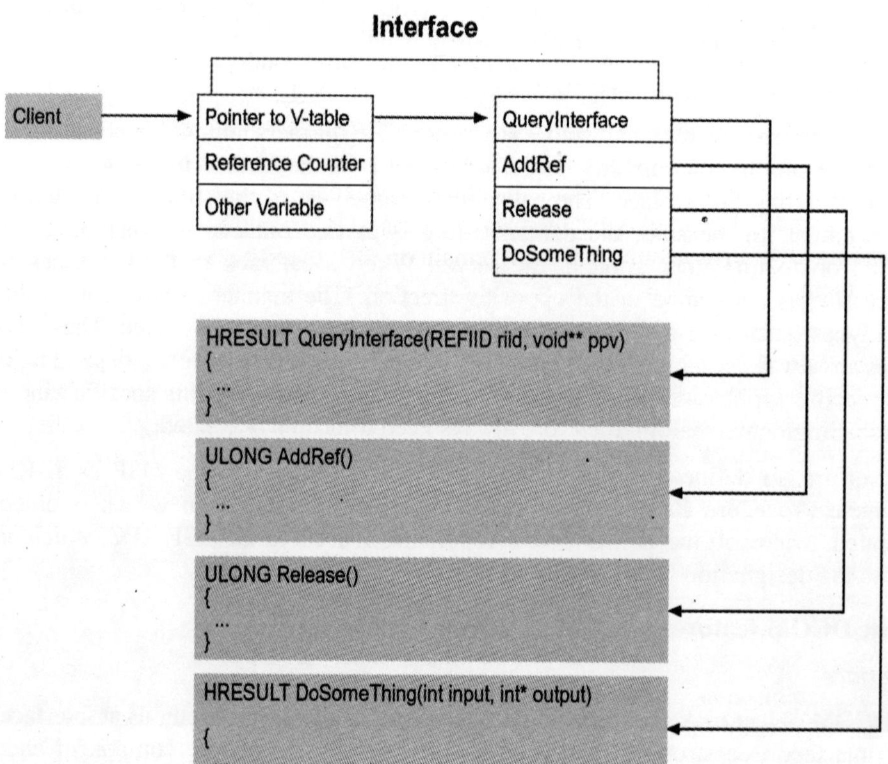

Figure 3.2: *Access to an interface*

The public member function of the class is declared as *pure virtual*. This means that the actual functionality is implemented somewhere else (pure) and that the function can be redefined (overloaded) in a derived class (virtual). The C++ compiler enters the address of the virtual functions in a special table – the v-table (virtual function table). This structure is used in C++ for implementing polymorphism. In DCOM it serves as the basis for accessing interfaces.

The object's memory area also contains a pointer to the v-table. This pointer is used by the client. **Figure 3.2** illustrates the different structures. Every programming environment that can create the corresponding structures can be used for implementing DCOM objects.

The functionality of the method at the specified interface (and other interfaces) is implemented in a COM class, for example, in C++. A component can contain one or several COM classes (DCOM objects). Every OPC server comprises several DCOM objects. For each DCOM interface, a separate v-table structure exists.

Proxy and stub

The existence and functionality of ORPC were mentioned above when explaining DCOM's roots. In DCOM, proxy and stub components are implemented in one DLL. But how is a proxy-stub-DLL created? An Interface Definition Language (IDL) similar to C exists for DCOM. This language contains a description of the COM class(es), (DCOM objects), and their interfaces with methods and parameters. All OPC specifications include the applicable interface description in their Appendix. Microsoft offers an IDL compiler which generates the corresponding C files from the IDL. These files, in turn, are compiled and linked into a DLL. The client and server components use this functionality at runtime. The advantage of this is that components can be implemented in any language. What all these components need to do is access a DLL. They do not have to implement the data exchange functionality beyond the process and PC boundaries by themselves. This also means, however, that the corresponding OPC Proxy/Stub DLL must remain on the PC as long as products exist which use them.

Object generation

The interfaces have been defined. The COM classes have been implemented. What remains to be done to complete the DCOM component? Both in the client and the server implementation, the DCOM runtime must be initialized before first use of the functionality and uninitialized before termination of the program. This is done by calling CoInitialize and CoUninitialize.

How does a client launch the server? To ensure local transparency, the DCOM object is not created in the server directly by the client (e.g. with the constructor new) but via a ClassFactory object, which is part of the server component and is in charge of creating the server's DCOM objects. In the client, the method CoCreateInstance is called. CoCreateInstance is a COM-API function. At first, the client will thus not interact directly with the component. If it did, no local transparency could be achieved. The DCOM runtime uses the CLSID and looks at the registry for the location of the binary file in which the COM class has been implemented. **Figure 2.1** shows the registry settings for an OPC server. The DCOM runtime then starts the corresponding program. When the DCOM runtime has been initialized by the started server component, CoRegisterClassObject is called. In the call, an interface pointer to the ClassFactory is passed to the runtime.

The ClassFactory object implements an interface IClassFactory with the methods CreateInstance and LockServer. The methods are easy to implement. In CreateInstance, the DCOM object is created with the constructor new, and the desired interface pointer is queried and returned. CreateInstance is called within CoCreateInstance. As a result of the CoCreateInstance call, the client receives an interface pointer to the DCOM object in the server. It can now call methods at this interface or use QueryInterface to query for other interface pointers.

Return values of methods

A method at an interface can deliver a return value which contains information on whether its execution was successful or has failed. The value is designated as HRESULT and has a length of 32 bits. Microsoft has already defined several bit patterns for errors that are related to the operating system or to the network. Further errors can be defined, as has been done in the OPC specifications.

Identifiers

This section will give a few details on the identifiers used for various tasks in DCOM. All of these identifiers are GUIDs (Globally Unique Identifiers). They have a length of 128 bits and are globally unique due to their generation. GUIDs are used, for example, as:

- CLSID – for the unique identification of a component. This value is used as a method parameter for CoCreateInstance for example.
- IID – for the unique identification of an interface. This value is used as a method parameter in QueryInterface for example.
- CATID – for the unique identification of component categories.

Creation of registry entries

Implementing the basic parts for simple DCOM servers and DCOM clients has now been completed. There is one last question: how is the required information (CLSID, IID, etc.) entered into the registry? There are two possibilities:

- Creating a registry file (*.rgs) that contains all the information. The entries are made when calling regedit with *.rgs. This is not very flexible and user-friendly, though. Another drawback is that information can only be entered but not deleted in this way.
- Self-registration. In this case, the server component itself enters the information in the registry by making use of corresponding API functions. This approach is more flexible as entries can be adjusted at runtime (e.g. reference to the current directory of the component). In addition, you can also implement the unregistering functionality with the API functions. In an OutProc implementation, the command line parameters –RegServer and –UnRegServer are used to invoke the described functionality. The corresponding functionality also needs to be implemented.

Using threads

The concepts discussed so far relate to the simpler part of DCOM. But there is also an area where things get a bit trickier. This area has to do with the use of threads.

Basically, threads are not DCOM-specific, not even Microsoft-specific. Threads are a programming aid used for running different functionality in parallel. When using threads, it is important to synchronize access to shared resources, such as global variables.

Within DCOM components, there are also many functions that need to be executed in parallel. A client can have different threads in which it interacts with different servers. A server uses different threads for processing the method calls at interfaces. DCOM assists the programmer in the synchronization.

Two different types of DCOM applications can be implemented:

- Applications with user interface.
- Applications without user interfaces.

Correspondingly, DCOM offers two threading models:

- Threading model for applications with user interfaces, in which the threads are synchronized using message queues. There are two reasons for this. Firstly, an application and the user interface generally interact via message queues. The queue can then also be used by DCOM. Secondly, DCOM's history is closely connected with message queues, due to its roots in DDE.
- Threading model for worker components. Here, DCOM does not provide any aids for ensuring proper synchronization.

How are these two models implemented? The basic unit of thread security in DCOM is an apartment, which means that access to apartments is synchronized by DCOM. The synchronization in the apartment is the task of the implementation.

In a single threaded apartment (STA), only one thread at a time can be executed. The programmer leaves the synchronization of parallel accesses up to DCOM. DCOM uses messages for this purpose, thus serializing the calling of methods. This solution reduces the performance as new access is blocked (probably for a long time). This process is secure, however, if components that do not support multi-threading are used.

In a multi threaded apartment (MTA), several threads can be executed at the same time. Programmers need to program the synchronization mechanisms themselves. Here, methods are called directly.

In one process, there can be none, one, or several STAs, and no or one MTA.

How are these threading models used? A component informs DCOM of the threading model when CoInitialize(Ex)[11] is called. InProc servers do not call CoInitialize if it

11 The various threading models were only created with the introduction of DCOM. An enhancement of the method became necessary to pass the supported threading model to the DCOM library during initialization. Therefore, a method CoInitializeEx exists that supports this parameter.

has already been called in this thread by the client. They indicate the supported threading model by an entry in the registry.

It is important to keep in mind that interface pointers are marshalled not only during the transition from one process to another but also from one apartment to another. Marshalling is omitted when client and server use the same threading model. To allow interoperability, an InProc server should always support both threading models. Direct method calls between client and server can then be carried out without involving the proxy/stub DLL. Otherwise, the interchange between the threading models will be performed via DCOM.

Setting up a callback connection

Servers can transmit information to the client via callback connections. The sequence during setup and use of the connection is illustrated in **Figure 3.3**. The client first calls the method FindConnectionPoint at the interface IConnectionPointContainer (1). At the corresponding ConnectionPoint, the client then calls the method Advise (2). To this method, the client passes a pointer to an interface of a DCOM object which is implemented by the client. The server can now call methods at the client object's interface (3).

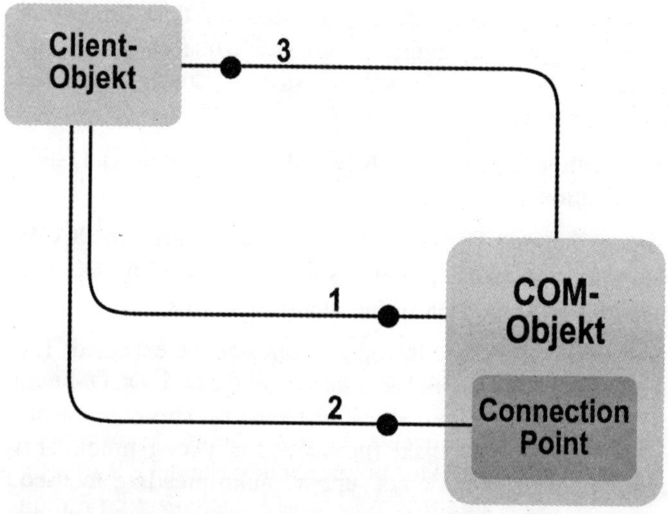

Figure 3.3: Setup and use of a callback connection

3.2.2 Implementing a DCOM server

This section describes the implementation of parts of a DCOM server to illustrate the DCOM basics dealt with previously in this chapter. This example is not intended as a basis for the development of OPC servers. Class libraries (MFC, ATL) are available for that purpose. These are not cited in the examples below, which demonstrate the basic principles only.

The implementation comprises the following functionality:

– Initializing and uninitializing the DCOM runtime

– Registering the object in the DCOM runtime

– CreateInstance and LockServer of the ClassFactory object

– Methods of the interfaces IOPCCommon and IOPCBrowseServerAddressSpace.

In addition to the other header files, the statements about the OPC interfaces also need to be included. The file OPCCmnServer.h contains the class definitions. The files opccomn.h and opcda.h are generated by the MIDL compiler from the IDL of the *OPC Common Definitions and Interfaces Specification* and the *OPC Data Access Specification*. They contain definitions for the interfaces of the specifications. The files opccomn_i.c and opcda_i.c contain the interface identifiers for the two specifications.

```
...
#include "opccomn.h"        // OPC common header
#include "opcda.h"          // OPC data access header
#include "initguid.h"
#include "opccomn_i.c"      // OPC common interface ids
#include "opcda_i.c"
#include "OPCCmnServer.h"
...
```

Further, you need to define several global variables. The global variables include, for example, the counters that manage the life cycle of the object and are taken into account in determining whether the component is to remain in the memory after its termination. The CLSID for the component needs to be defined also.

```
// object count
ULONG g_objCnt = 0;
// lock count
ULONG g_lockCnt = 0;
// class id of the server object
// {381B2F22-C0FB-4948-B680-918A3B6F5D1A}
CLSID g_clsid = { 0x381b2f22, 0xc0fb, 0x4948,
    { 0xb6, 0x80, 0x91, 0x8a, 0x3b, 0x6f, 0x5d, 0x1a } };
// end event of the application
HANDLE g_endEvent = INVALID_HANDLE_VALUE;
```

The server presented here has a very plain design, so that a simple main program (main) is sufficient. You first declare an interface pointer and create the ClassFactory object. Next, you initialize the DCOM runtime. As you are implementing a multithreaded application, you need to use CoInitializeEx and pass the corresponding threading model as the input parameter. By calling CoRegisterClassObject, the ClassFactory passes information about itself to the DCOM runtime. The application now continues running and waits for the end event. As soon as the end event is received, the ClassFactory informs the DCOM runtime that it is no longer available. The DCOM runtime is then uninitialized.

```
int main(int argc, char* argv[])
{
    CClassFactory *pFactory = new CClassFactory();
    IUnknown *pICF;
    DWORD registerId;
```

```
    // initialize COM libraries
    CoInitializeEx(NULL, COINIT_MULTITHREADED);
    // create end event
    g_endEvent = CreateEvent(NULL, FALSE, FALSE, NULL);
    // register class factory
    pFactory->QueryInterface(IID_IUnknown, (void **)&pICF);
    if (SUCCEEDED(CoRegisterClassObject(g_clsid, pICF, CLSCTX_LOCAL_SERVER,
        SREGCLS_MULTIPLEUSE, &registerId)))
    {
        pICF->Release();        // release reference to registered interface
        // wait for the end of the application
        WaitForSingleObject(g_endEvent, INFINITE);
        // revoke class factory
        CoRevokeClassObject(registerId);
    }
    // free allocated resources
    CloseHandle(g_endEvent);
    // uninitialize COM libraries
    CoUninitialize();
    mreturn 0;
} // main
```

What does the implementation of the component's functionality comprise?

The COM Class CServer has a constructor and a destructor. Here both are kept simple, as this example provides only few functions.

```
//————————————————————————————————
// CServer
//————————————————————————————————
CServer::CServer(void)
{
    m_refCnt = 0;
    g_objCnt++;
    m_localeID = MAKELCID(LANG_SYSTEM_DEFAULT, SORT_DEFAULT);
    m_clientName = wcsdup(L"");
    m_curBrowsePath = wcsdup(L"");
}
CServer::~CServer(void)
{
    free((void *)m_clientName);
    free((void *)m_curBrowsePath);
    g_objCnt–;
}
```

The next lines show an implementation of QueryInterface. This method is called by the client. The implementation of this client will be described in the next section. The COM class has three interfaces – IUnknown, IOPCCommon, and IOPCBrowseServerAddressSpace. By passing an IID, the client can query for an interface pointer. When the client has passed the IID of an implemented interface, it receives the pointer to the interface as a result. A typecast ensures that the client receives exactly the interface it has queried. If the client passes the IID of an interface that is not supported, E_NOINTERFACE is returned as the result. According to the DCOM rules for managing the life cycle of the objects, AddRef is called by the server before the interface pointer is returned in order to increment the object's global counter.

```
//------------------------------------------------------------
//  QueryInterface
//  returns:
//      S_OK - interface returned
//      E_NOINTERFACE - interface not supported
STDMETHODIMP CServer::QueryInterface(
    IN REFIID riid,             // id of the requested interface
    OUT LPVOID *pIf)            // pointer to the interface
{
        *pIf = NULL;
        // the object supports the IUnknown and the IOPCCommon interfaces
        if ((riid == IID_IUnknown) || (riid == IID_IOPCCommon))
        *pIf = (IOPCCommon *)this;
    if (riid == IID_IOPCBrowseServerAddressSpace)
        *pIf = (IOPCBrowseServerAddressSpace *)this;
    if (*pIf == NULL)
        return E_NOINTERFACE;    // interface not supported
    AddRef();                    // increment reference count of the object
    return S_OK;
} // QueryInterface
```

Below, you can see the implementation of the AddRef method. The reference counter for the object is incremented before the pointer is returned.

```
//------------------------------------------------------------
// AddRef
// returns:
//      the new reference count of the object
//
STDMETHODIMP_ (ULONG) CServer::AddRef(void)
{
    return ++m_refCnt;
} // AddRef
```

The counterpart to AddRef is Release. Here, the reference counter is decremented. If the counter equals 0, the destructor for the object is called. This happens when the client releases the last interface pointer to an object it has created. If no further objects exist (g_objCnt = 0), the end event is signaled to the main thread of the application. This occurs if no further clients are accessing the component. On receipt of this event, the application terminates itself as implemented above.

```
//------------------------------------------------------------
// Release
// returns:
// the new reference count of the object
//
STDMETHODIMP_ (ULONG) CServer::Release(void)
{
    if(-m_refCnt != 0)
        return m_refCnt;
    delete this;
    // end the application if the last object was released
    if (g_objCnt == 0)
        SetEvent(g_endEvent);
    return 0;
} // Release
```

The ClassFactory will be described before proceeding with the implementation of the OPC functionality.

The following section implements the ClassFactory functionality. It comprises a constructor and, of course, QueryInterface, AddRef, and Release. They are implemented in the same way as the implementation shown above, except that now, other IIDs serve as the basis of decision in QueryInterface.

```
//————————————————————————————————————
// CClassFactory |
//————————————————————————————————————
CClassFactory::CClassFactory(void)
{
    m_refCnt = 0;
}
```

The CreateInstance method is of even greater importance. When implementing CreateInstance, an object of the COM Class CServer is created with the constructor. The IID for the interface for which a pointer is to be returned is one of the input parameters of CreateInstance. CreateInstance is called by the DCOM runtime during the execution of CoCreateInstance. The client passes an IID when this method is called. This is exactly the identifier taken into account again here.

```
//————————————————————————————————————
// CreateInstance
// returns:
//      S_OK - object created
//   E_NOINTERFACE  - interface not supported
//   CLASS_E_NOAGGREGATION  - no aggregation supported
//
STDMETHODIMP CClassFactory::CreateInstance(
    IN LPUNKNOWN pUnkOuter,   // outer interface of the object (needed for
                                                               aggregation)
    IN REFIID riid,             // id of the requested interface
    OUT LPVOID *pIf)            // pointer to the interface
{
    HRESULT res;
    *pIf = NULL;
    // no aggregation supported
    if (pUnkOuter != NULL)
        return CLASS_E_NOAGGREGATION;
    // create object
    CServer *pServer = new CServer();
    if (SUCCEEDED(pServer->QueryInterface(riid, pIf)))
        res = S_OK;
    else
    {
        delete pServer;
        res = E_NOINTERFACE;
    }
    return res;
} // CreateInstance
```

Once the DCOM object has been created, the client can access methods at interfaces. In the following, only the implementation of a method at the IOPCCommon interface will be explained at first.

The method SetLocaleID is a method at this interface. It can be called by the client on receipt of the pointer to this interface. By using this method, the client can set the language (LocaleId) to be used subsequently. If the language is not supported, the error code E_INVALIDARG is returned. At this point, less can actually be seen of DCOM.

```
//------------------------------------------------------------
// SetLocaleID (interface IOPCCommon)
//
//   - set locale id of server
//
// returns:
//   S_OK   - set locale id
//   E_INVALIDARG   - invalid locale id for this server
//
STDMETHODIMP CServer::SetLocaleID(
    IN LCID lcid)           // locale id
{
    HRESULT res;
    switch (lcid)
    {
        // supported locale ids
        case MAKELCID(LANG_SYSTEM_DEFAULT, SORT_DEFAULT):
        case MAKELCID(LANG_USER_DEFAULT, SORT_DEFAULT):
        case MAKELCID(MAKELANGID(LANG_GERMAN, SUBLANG_GERMAN),
             SORT_DEFAULT):
            m_localeID = lcid;
            res = S_OK;
        break;
        default:
            res = E_INVALIDARG;
        break;
    }
    return res;
} // SetLocaleID
```

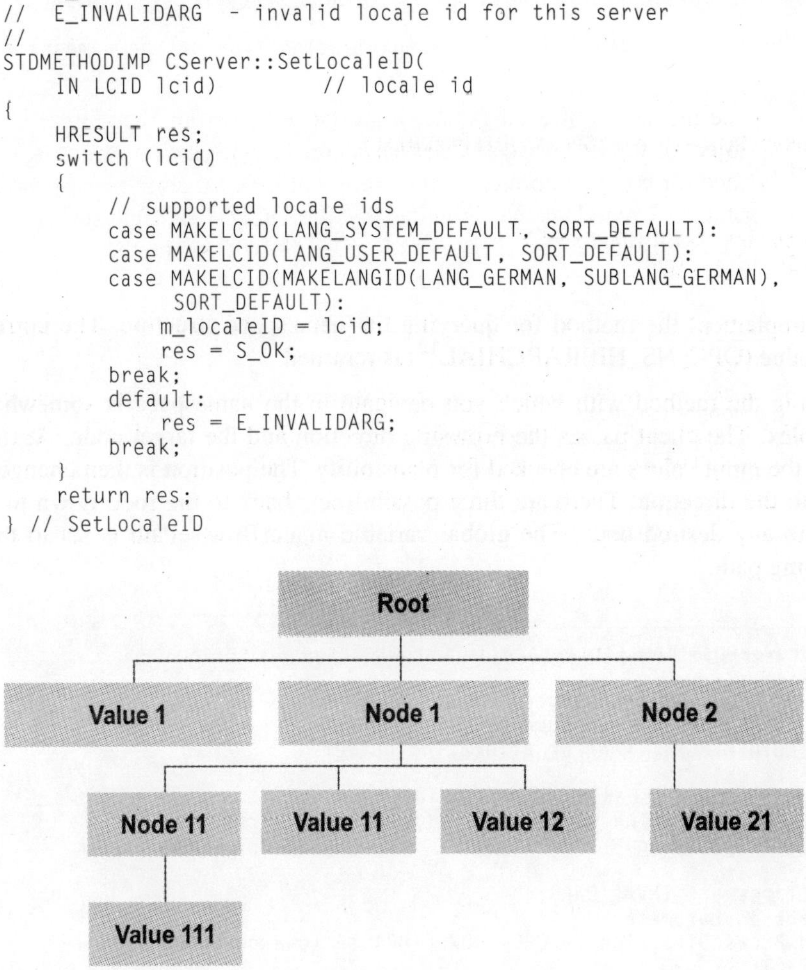

***Figure 3.4:** Namespace of the sample server*

This section deals with the methods at the interface used for browsing the namespace. After you have read the section on the *Data Access Specification*, you may be interested in knowing how the client browses the namespace. In the following section, the corresponding client will be described. The namespace is structured hierarchically and illustrated in **Figure 3.4**.

The IOPCBrowseServerAddressSpace interface provides five methods, which are implemented as shown below.

```
//————————————————————————————————————
// QueryOrganization (interface IOPCBrowseServerAddressSpace)
// returns:
//      S_OK   - returned type
//      E_INVALIDARG   - bad arguments
//
STDMETHODIMP CServer::QueryOrganization(OUT OPCNAMESPACETYPE *pNameSpace-
Type)
                                         //organization type of the namespace
{
    if (pNameSpaceType)
    {
        *pNameSpaceType = OPC_NS_HIERARCHIAL;
        return S_OK;
    }
    else
        return E_INVALIDARG;
} // QueryOrganization
```

First, you implement the method for querying the namespace structure. The corresponding value (OPC_NS_HIERARCHIAL[12]) is returned.

Implementing the method with which you navigate in the namespace is somewhat more complex. The client passes the browsing direction and the target node. At the beginning, the input values are checked for plausibility. The position is then changed according to the direction. There are three possibilities: back to the root, down to a node, and to any desired node. The global variable m_curBrowsePath is set to the corresponding path.

```
//————————————————————————————————————
// ChangeBrowsePosition (interface IOPCBrowseServerAddressSpace)
// returns:
//      S_OK   - position changed
//      E_INVALIDARG   - bad name or direction
//      E_OUTOFMEMORY   - error allocating buffer
//
STDMETHODIMP CServer::ChangeBrowsePosition(
    IN OPCBROWSEDIRECTION dwBrowseDirection,      // direction
    IN LPCWSTR szString)                           // down node name
{
    HRESULT res = E_INVALIDARG;
    // check in parameter
    if ((dwBrowseDirection != OPC_BROWSE_UP) && (dwBrowseDirection !=
OPC_BROWSE_TO) &&
        (dwBrowseDirection != OPC_BROWSE_DOWN))
        return E_INVALIDARG;
    if ((dwBrowseDirection == OPC_BROWSE_DOWN) && ((szString == NULL) ||
        (wcslen(szString) == 0)))
        return E_INVALIDARG;
```

12 This is not a spelling error in this book but in the specification.

```
    if ((dwBrowseDirection == OPC_BROWSE_UP) && (wcscmp(m_curBrowsePath,
L"") == 0))
        return E_INVALIDARG;
    switch (dwBrowseDirection)
    {
        case OPC_BROWSE_UP:
        {
            LPCWSTR upperLevel = NULL;
            LPCWSTR name = NULL;
            reducePath(m_curBrowsePath, FALSE, upperLevel, name);
            if (!upperLevel)
            { // reached the root object
                m_curBrowsePath[0] = L'';
            }
            res = S_OK;
        }
        break;
        case OPC_BROWSE_DOWN:
        {
            LPWSTR path = wcsdup(m_curBrowsePath);
            if (isValidBranch(m_curBrowsePath, szString))
            { // found valid branch
                LPWSTR path =
(LPWSTR)malloc((wcslen(m_curBrowsePath) + wcslen(szString) + 2) * 2);
                if (path)
                {
                    if (wcslen(m_curBrowsePath))
                    {
                        wcscpy(path, m_curBrowsePath);
                        wcscat(path, L".");
                    }
                    else
                        path[0] = L'\0';
                    wcscat(path, szString);
                    free(m_curBrowsePath);
                    m_curBrowsePath = path;
                    res = S_OK;
                }
                else
                    res = E_OUTOFMEMORY;
            }
            else
                res = E_INVALIDARG;
        }
        break;
        case OPC_BROWSE_TO:
        {
            LPWSTR path = wcsdup(szString);
            LPCWSTR upperLevel = NULL;
            LPCWSTR name = NULL;
            reducePath(path, FALSE, upperLevel, name);
            if ((wcscmp(szString, L"") == 0) ||
            (isValidBranch(upperLevel, name)))
            {
                free(m_curBrowsePath);
                m_curBrowsePath = wcsdup(szString);
                res = S_OK;
            }
            else
                res = E_INVALIDARG;
            free(path);
        }
        break;
```

```
    }
    return res;
} // ChangeBrowsePosition
```

Implementing the method with which the client queries the ItemId (identifier) is even more complex. By passing the corresponding values as parameters, the client determines whether:

- it wants to receive nodes or leaves of the namespace
- it wants the results to be filtered with regard to the VARIANT data type, access rights, or text contained

At the beginning, the input parameters are again checked for plausibility. For reasons of simplicity, no filtering will take place in the example. You will find an example of string filtering in the *Data Access Specification.*

The first decision criterion is the current value of m_curBrowsePath. The pointer of the client can be located either at the root of the namespace or at any other node. The ItemId is then returned for each of these nodes according to the value of BrowseFilterType. The client can request leaves or nodes (branches) or all information below the current position (flat). The result is saved in a string enumerator. String enumerators are also DCOM objects and, therefore, the interface pointer to the object is transmitted to the client. This solution optimizes the transmission of information: the client receives only an interface pointer and not a set of strings.

```
//——————————————————————————————————————
// BrowseOPCItemIDs (interface IOPCBrowseServerAddressSpace)
// returns:
//   S_OK   - string enumerator returned
//   S_FALSE   - empty string enumerator returned
//   E_INVALIDARG   - bad argument
//
STDMETHODIMP CServer::BrowseOPCItemIDs(
    IN OPCBROWSETYPE dwBrowseFilterType,      // get what? tags or nodes
    IN LPCWSTR szFilterCriteria,              // string filter
    IN VARTYPE vtDataTypeFilter,              // datatype filter
    IN DWORD dwAccessRightsFilter,            // access rights filter
    OUT LPENUMSTRING *ppIEnumString)          // child names
{
    // check in and out parameter
    if (!ppIEnumString)
        return E_INVALIDARG;
    *ppIEnumString = NULL;
    if ((dwBrowseFilterType != OPC_BRANCH) &&
    if ((dwBrowseFilterType != OPC_LEAF) &&
    if ((dwBrowseFilterType != OPC_FLAT))
        return E_INVALIDARG;
    // this implementation ignores the filters!
    DWORD idx = 0;
    CEnumString *stringEnum = new CEnumString();
    if (wcscmp(m_curBrowsePath, L"") == 0)
    { // root branch
        switch (dwBrowseFilterType)
        {
            case OPC_LEAF:
                stringEnum->addString(L"Value 1", idx++);
            break;
```

```
        case OPC_BRANCH:
            stringEnum->addString(L"Node 1", idx++);
            stringEnum->addString(L"Node 2", idx++);
        break;
        case OPC_FLAT:
            stringEnum->addString(L"Value 1", idx++);
            stringEnum->addString(L"Node 1.Value 11", idx++);
            stringEnum->addString(L"Node 1.Value 12", idx++);
            stringEnum->addString(L"Node 1.Node 11.Value 111", idx++);
            stringEnum->addString(L"Node 2.Value 21", idx++);
        break;
    }
}
else if ((wcscmp(m_curBrowsePath, L"Node 1.Node 11") == 0) ||
    (wcscmp(m_curBrowsePath, L"Node 1/Node 11") == 0) ||
    (wcscmp(m_curBrowsePath, L"Node 1\\Node 11") == 0))
{
    switch (dwBrowseFilterType)
    {
        case OPC_LEAF:
        case OPC_FLAT:
            stringEnum->addString(L"Value 111", idx++);
        break;
    }
}
else if (wcscmp(m_curBrowsePath, L"Node 1") == 0)
{
    switch (dwBrowseFilterType)
    {
        case OPC_LEAF:
            stringEnum->addString(L"Value 11", idx++);
            stringEnum->addString(L"Value 12", idx++);
        break;
        case OPC_BRANCH:
            stringEnum->addString(L"Node 11", idx++);
        break;
        case OPC_FLAT:
            stringEnum->addString(L"Value 11", idx++);
            stringEnum->addString(L"Value 12", idx++);
            stringEnum->addString(L"Node 11. Value 111", idx++);
        break;
    }
}
else if (wcscmp(m_curBrowsePath, L"Node 2") == 0)
{
    switch (dwBrowseFilterType)
    {
        case OPC_LEAF:
        case OPC_FLAT:
            stringEnum->addString(L"Value 21", idx++);
        break;
    }
}
stringEnum->QueryInterface(IID_IEnumString, (void **)ppIEnumString);
return (idx > 0) ? S_OK : S_FALSE;
} // BrowseOPCItemIDs
```

In contrast, implementing the GetItemId method is less complex. At the beginning, the input values are checked for plausibility. The auxiliary function isValidItemId checks whether the ItemId passed as a parameter exists in the namespace. Then the fully qualified ItemId is generated, in which the delimiters are inserted.

```
//———————————————————————————————
// GetItemID (interface IOPCBrowseServerAddressSpace)
//
//  - get the full qualified path for a tag or node name
//
// returns:
//  S_OK                  - returned fully qualified name
//  E_INVALIDARG          - return pointer was NULL
//  E_OUTOFMEMORY         - error allocating buffer for return
//
STDMETHODIMP CServer::GetItemID(
    IN LPWSTR szItemDataID,  // element name
    OUT LPWSTR *szItemID)    // fully qualified name
{
    // check in and out parameter
    if (!szItemID)
        return E_INVALIDARG;
    *szItemID = NULL;
    if (!szItemDataID)
        return E_INVALIDARG;
    // got valid item from the client ?
    if (isValidItemID(m_curBrowsePath, szItemDataID))
    { // yes  - copy it into the return parameter
        *szItemID =
        (LPWSTR)CoTaskMemAlloc((wcslen(m_curBrowsePath) +
                                wcslen(szItemDataID) + 2) * 2);
        if (*szItemID)
        {
            // construct the item id
            if (wcslen(m_curBrowsePath))
            {
                wcscpy(*szItemID, m_curBrowsePath);
                wcscat(*szItemID, L".");
            }
            else
                (*szItemID)[0] = L'\0';
            wcscat(*szItemID, szItemDataID);
            return S_OK;
        }
        else
        { // no memory for the return
            return E_OUTOFMEMORY;
        }
    }
    return E_INVALIDARG;
} // GetItemID
```

The namespace in this example does not support AccessPath. The implementation of this method is therefore not explained.

The implementation uses auxiliary functions which are described in **Table 3.1.**

Table 3.1: *Auxiliary functions of the sample server*

isValidItemID	Checks whether there is a valid representation of the input parameter of the method GetItemId
isValidBranch	Checks whether the input parameter of ChangeBrowsePosition points to an existing node of the namespace.
reducePath	Used to find the separator for namespace paths.

The implementation of the sample server has now been completed.

3.2.3 Implementing a DCOM client

This section deals with the implementation of parts of a DCOM client. The client to be implemented in this section will access interfaces and methods of the server created in the previous section.

The include files with information about the OPC interfaces need to be integrated for this application also.

```
...
#include "opccomn.h"     // OPC common header
#include "opcda.h"       // OPC data access header
#include "initguid.h"
#include "opccomn_i.c"   // OPC common interface ids
#include "opcda_i.c"     // OPC data access interface ids
...
```

Only one global variable – CLSID – needs to be defined for this example implementation. It is the same CLSID that was used for implementing the server. This is necessary as the client should access this server. The client can query for the CLSID by using functionality of the OPCServerBrowser (see paragraph 2.2.2).

```
CLSID g_clsid = { 0x381b2f22, 0xc0fb, 0x4948,
 { 0xb6, 0x80, 0x91, 0x8a, 0x3b, 0x6f, 0x5d, 0x1a } };
```

Further, the browseBranch method is declared. This method implements access to the namespace.

```
void browseBranch(IN IOPCBrowseServerAddressSpace *pBrowse);
```

The client also has a very simple design. The implementation of the application is therefore located in a simple main routine. Since the client of course also uses DCOM functionality, the DCOM runtime is initialized right at the beginning. It is then followed by the call of CoCreateInstance.

```
int main(int argc, char* argv[])
{
    IOPCCommon *pServer;      // COM interface of the server
    IOPCBrowseServerAddressSpace *pBrowse;
                              // Browse COM interface of the server
    DWORD lcid;
    // initialize COM libraries
    CoInitializeEx(NULL, COINIT_MULTITHREADED);
    // create an instance of the server object
    if (SUCCEEDED(CoCreateInstance(g_clsid, NULL, CLSCTX_LOCAL_SERVER,
                                   IID_IOPCCommon, (void **)&pServer)))
...
```

Once the server has been successfully launched, the existing functionality can now be accessed. This is done simply by invoking methods at the interface. After all methods have been executed at the IOPCCommon interface, QueryInterface is called for querying the pointer to the browser interface. The browse position is set to the root of the namespace. Then browseBranch is called.

```
{ //call some functions on the interface
    printf(„set client name to OPCCmnClient\n");
    pServer->SetClientName(L"OPCCmnClient");
    pServer->GetLocaleID(&lcid);
    printf(„got default LCID 0x%8.8X\n", lcid);
    printf(„set LCID to german (0x407)\n");
    pServer->SetLocaleID(MAKELCID(MAKELANGID (LANG_GERMAN, SUBLANG_GERMAN),
                                                              SORT_DEFAULT));
    pServer->GetLocaleID(&lcid);
    printf(„got new LCID 0x%8.8X\n", lcid);
   if (SUCCEEDED(pServer->QueryInterface(IID_IOPC BrowseServerAddressSpace,
                                                           (void**)&pBrowse)))
        {
            wprintf(L"\n\n- Browse -\n");
            pBrowse->ChangeBrowsePosition(OPC_BROWSE_TO, L"");
            browseBranch(pBrowse);
            // release the interface
            pBrowse->Release();
        }
        // release the server object
        pServer->Release();
    }
    else
    {
        wprintf(L"can't create server object  - check the registry!\n");
    }
```

As soon as the method for browsing the namespace has been executed, the reference counter for this interface is decremented and the pointer is released. The same happens with the pointer to IOPCCommon. Since this interface pointer is the last the client has used for the server's object, this call will result in the example server terminating itself.

At the end, the client needs to release the DCOM library. This is done by calling CoUninitialize.

```
    // uninitialize COM libraries
    CoUninitialize();
    return 0;
} // main
```

The browseBranch function first returns the identifier of the leaves of the current browse position. The function is then called recursively for all leaves.

```
//————————————————————————————————
// browseBranch
//
// - browse the current branch and its subtrees
//
void browseBranch(
    IN IOPCBrowseServerAddressSpace *pBrowse)      // browse interface
    {
    IEnumString *stringEnum;
    LPWSTR str;
    LPWSTR id;
    ULONG fetched;
    // print the leafs
    if (SUCCEEDED(pBrowse->BrowseOPCItemIDs(OPC_LEAF, L"*", VT_EMPTY, 0,
                                                         &stringEnum)))
```

```
    {
        while ((stringEnum->Next(1, &str, &fetched) == S_OK))
        {
            if (SUCCEEDED(pBrowse->GetItemID(str, &id)))
            {
                wprintf(L"%s\n", id);
                CoTaskMemFree(id);
                CoTaskMemFree(str);
            }
        }
        stringEnum->Release();
    }
    // print the branches
    if (SUCCEEDED(pBrowse->BrowseOPCItemIDs(OPC_BRANCH, L"*", VT_EMPTY, 0,
                                                              &stringEnum)))
    {
        while ((stringEnum->Next(1, &str, &fetched) == S_OK))
        {
            if (SUCCEEDED(pBrowse->ChangeBrowsePosition (OPC_BROWSE_DOWN,
                                                              str)))
            {
                browseBranch(pBrowse);
                pBrowse->ChangeBrowsePosition(OPC_ BROWSE_UP, L"");
            }
        }
        stringEnum->Release();
    }
} // browseBranch
```

3.3 .NET

.NET is based on a Common Language Runtime (CLR) which is located above an operating system layer and which can also be ported to non-Windows operating systems. The CLR has several important characteristics, on which other parts of the framework are based, and which enable each .NET language to access components written in other .NET languages at design time and runtime. **Figure 3.5** shows the components.

These characteristics include a common type system and common base classes, and definition and use of an intermediate code (MIL – Microsoft Intermediate Language). Compilers can generate this intermediate code; thus, components in .NET environments can be processed in any type of architecture. Naturally, the possibility of generating and running components in the native code exists as well. In addition to the executable code, information on the component itself is also stored in a standardized manner. Thus, no additional files are necessary for accessing the component. These meta-data are automatically created during compilation and are therefore always up to date.

The next layer, i.e. the Services Framework, offers a set of services for specific areas (for example, base classes and programming support).

.NET supports two different types of applications – Web applications (Web Forms and Web Services) and Windows applications (Win Forms). Furthermore, in terms of Web applications, it is possible to distinguish between Web applications in direct connection with Web servers and Web browsers (Web Forms) and Web services independent of Web servers and browsers.

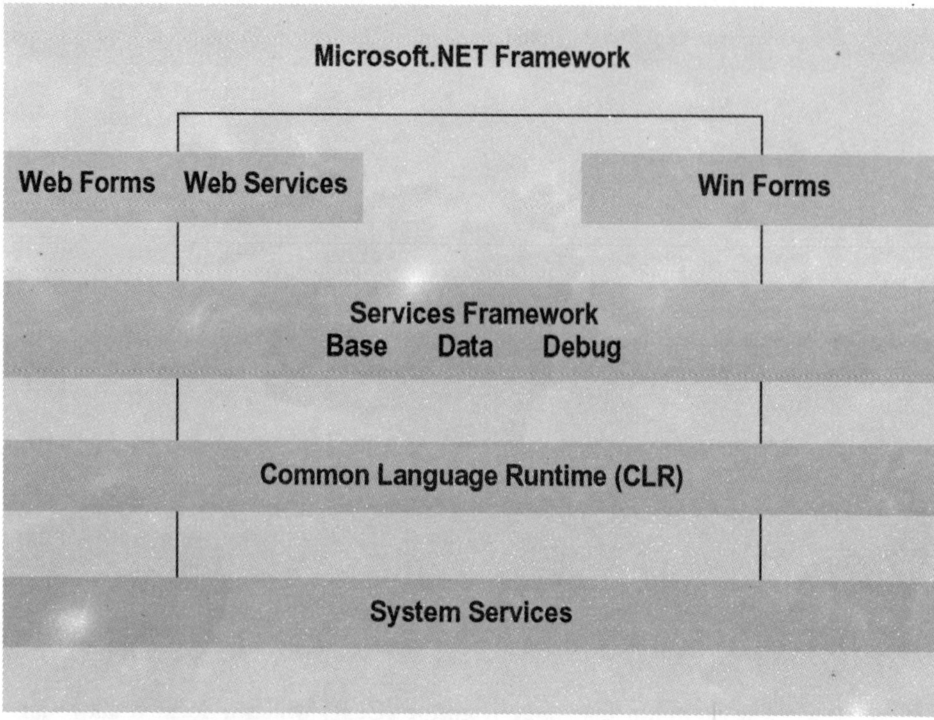

Figure 3.5: *Components of the .NET framework*

Not shown in **Figure 3.5**, but also of great importance, are the different forms of interaction between .NET components. .NET components can exchange information either by SOAP (Simple Object Access Protocol, XML + HTTP) or with binary encoding via TCP. Thus there is a type of firewall-friendly communication via HTTP and XML, and a highly performant type of communication with binary encoding usable in the Intranet. Components running in environments that do not support .NET can access .NET components via SOAP (XML + HTTP). The exchange between .NET components and DCOM components takes place via binary encoding of the data stream, based on traditional DCOM.

.NET in comparison with DCOM

Microsoft launches .NET also as the successor of DCOM, which is specifically targeted at use in the Internet. In what respect is .NET better than DCOM?

1. The development of the components is simplified, as the sophisticated particularities of DCOM (e.g. the implementation of interfaces and management of the life cycle) no longer apply.
2. Due to the availability of the uniform type system as well as uniform classes and meta-data, all languages supporting .NET can directly access .NET components.

This is not the case with DCOM; for example, programming access to a server in C++ is different from doing so in Visual Basic.

3. The uniform runtime environment (CLR) can be ported to other systems, thus developed components can run on these systems as well.
4. .NET uses SOAP as an interaction protocol between components, and easy integration in Internet applications is possible. These applications can then contain both .NET components and other components using SOAP for interaction.
5. Whereas DCOM is a closed Microsoft technology, .NET specifications are freely accessible and used for portation even today.

The impact of .NET on the future development of OPC

In examining this issue, it is necessary to distinguish two aspects:

1. future development of OPC products
2. future use of existing OPC products

As Microsoft regards .NET as the future development environment and is not developing DCOM further, the OPC Foundation must give thought to defining .NET interfaces for the existing specifications. However, this does not apply to the runtime environment. Here, DCOM will also be supported in the future so that existing OPC DCOM clients and servers can continue to be used.

Nevertheless, for existing clients and servers to work together with .NET components, a component that maps between the .NET model and DCOM must be employed. The interface of this component must be identical with the general interface for .NET OPC servers, which has yet to be defined. It should in the future be transparent for a .NET OPC client, whether it interacts with a DCOM OPC server or a .NET OPC server.

Figure 3.6 shows code lines for the implementation of such a component and the access to it from different .NET languages. C# (pronounced C sharp) is a new language developed by Microsoft in connection with .NET.

A component maps DCOM interfaces to .NET objects. The figure represents this process for the AddGroup method available at the OPCServer object. The different .NET languages access the object in the same manner. There are differences in terms of the languages' properties (e.g. variable declaration), but not in terms of object access.

Why is it useful to develop OPC servers and clients on the basis of .NET technology in the future, and what will change?

The development of products will be greatly simplified in terms of the implementation of the OPC interface. From different languages, the objects can be directly accessed. Servers can be written in different languages (e.g. C++, C#, VisualBasic.NET).

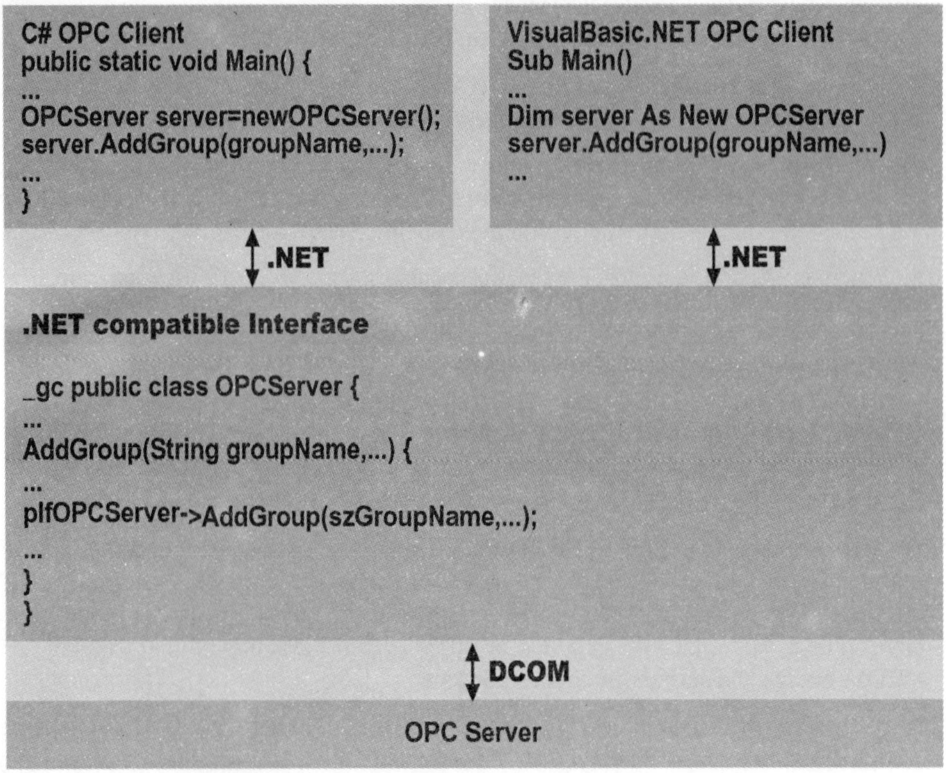

Figure 3.6: *Access to OPC-DCOM servers with .NET components*

Usage of .NET in future OPC products

A far more important reason for using .NET in future OPC products, however, is the implicit Internet capability of Web services. As already mentioned, .NET uses SOAP as an interaction protocol; SOAP uses XML to describe the method signatures and HTTP as the transport protocol. Thus, client and server solutions can be integrated in OPC applications supporting XML and HTTP. This support is available on very different systems; it can often be added very easily. For an OPC server with a .NET interface implemented as a Web service, implicit use for an Internet application would thus be possible immediately.

Outlook

The possibility of .NET portation has been addressed several times. A case in point is the Mono project (http://www.go-mono.com), an initiative that develops a Linux-based Open Source version of .NET. It should then be possible to run components developed on other platforms in this environment, or to develop them directly for this environment.

What would be the advantages to the user?

– platform-independent interoperability

– transparent data access from any location

For the moment, the OPC Foundation has taken the first step towards the use of .NET technology.

3.4 XML, SOAP and Web Services

There are many products that implement DCOM based OPC specifications. A few limitations apply, however, which need to be considered when using and developing these products:

- DCOM cannot be used across firewalls. PCs cannot be addressed directly across the firewall. But this is exactly what DCOM needs in order to perform an internal check. There is the possibility of transmitting DCOM over the Internet via Virtual Private Networks, but this involves extensive administrative work and even then is not always possible. This is due to both technical (different VPN protocols) and administrative reasons (VPN applies more to PCs than to applications).
- Some devices and applications that provide or require data do not run on Microsoft systems. This concerns applications in the ERP and MES areas as data consumers, and embedded devices as data sources.

In view of these limitations, the OPC Foundation started working on the OPC XML-DA specification, which is no longer based on DCOM, but on a technology that is independent of a specific operating system. The relevant components of this technology will therefore be described in greater detail in this section.

The eXtensible Markup Language (XML) is a flexible, easy to understand, and easy to learn data description language. Information is exchanged using readable XML documents that contain plain text with additional structuring information. An XML document contains elements and attributes. Elements can include further elements. Attributes are allocated to elements. The information is structured by the elements and attributes selected. XML determines that elements are characterized by freely definable tags. An XML document is called well-formed if it corresponds to the XML syntax. A document is called valid if, in addition, it corresponds to a model, the schema. The schema itself also uses the XML notation. A schema thus defines a language for structuring information in a specific application area. Preparation of XML documents and schemas, as well as validation and processing of files, is supported by a large number of tools. Processing of documents, which means their conversion, processing, and representation, can be controlled flexibly.

Applications can exchange XML documents in different ways, e.g. via floppy disk, e-mail, HTTP or TCP/IP. Practically all systems meanwhile support XML, which also allows heterogeneous systems to interact very easily by exchanging XML documents.

The Simple Object Access Protocol (SOAP) is an interaction protocol that links two technologies: XML and HTTP. HTTP is used as transport protocol. The parameters of the interactions are described with XML. SOAP is thus particularly predestined for the Internet. SOAP is a protocol independent of object architectures (DCOM, CORBA). A SOAP telegram consists of a header describing the structure of the HTTP call (Request/Response, host/PC, content type, and content length). This part is included in all the HTTP telegrams. A UniversalResourceIdentifier (URI) was added, which defines the end point/program and the method to be called. The method parameters are transferred as XML. The programmer is responsible for mapping the SOAP protocol to a concrete implementation. In the mean time, SOAP has been submitted to the W3C (World Wide Web Consortium) for standardization. In this context the name has changed to XML protocol. Version 1.2 is available since December 2001.

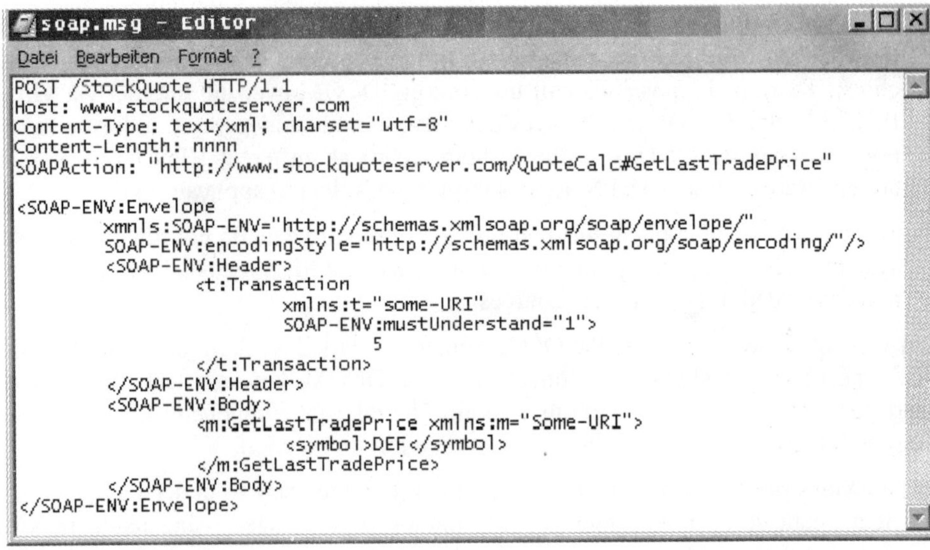


```
POST /StockQuote HTTP/1.1
Host: www.stockquoteserver.com
Content-Type: text/xml; charset="utf-8"
Content-Length: nnnn
SOAPAction: "http://www.stockquoteserver.com/QuoteCalc#GetLastTradePrice"

<SOAP-ENV:Envelope
        xmnls:SOAP-ENV="http://schemas.xmlsoap.org/soap/envelope/"
        SOAP-ENV:encodingStyle="http://schemas.xmlsoap.org/soap/encoding/"/>
        <SOAP-ENV:Header>
                <t:Transaction
                        xmlns:t="some-URI"
                        SOAP-ENV:mustUnderstand="1">
                                5
                </t:Transaction>
        </SOAP-ENV:Header>
        <SOAP-ENV:Body>
                <m:GetLastTradePrice xmlns:m="Some-URI">
                        <symbol>DEF</symbol>
                </m:GetLastTradePrice>
        </SOAP-ENV:Body>
</SOAP-ENV:Envelope>
```

Figure 3.7: SOAP request

Figure 3.7 shows a SOAP request transmitted by HTTP post. The first lines belong to the HTTP header, which is followed by an XML document. The request is addressed to the host www.stockquoteserver.com, where the GetLastTradePrice method is to be called in the QuoteCalc program.

The SOAP message comprises a header and a body. The header attributes describe how the request is to be processed. The body contains the method parameters: the symbol of the company for which the stock price is to be inquired.

The technologies presented so far in this section already allow the implementation of distributed applications that interact via SOAP, independently of the operating system and computer platform. But there is still something missing: A way to describe the interface provided by an application as well as a way to use this description to generate program sections which not only interact with the existing infrastructure (HTTP, etc.), but can be integrated into existing programs. This possibility is offered by Web services.

The programmatic interfaces made available by application components are referred to as Web Services. SOAP is used as interaction protocol between components.

Web Services are described using XML. The language used is WSDL (Web Services Description Language) and is standardized in the W3C. A schema defines the syntax and semantics of WSDL; based on this definition, schema for concrete interfaces are defined. This can be done by individual manufacturers or by an organization like the OPC Foundation. An application interacting with the Web Service will deliver valid XML messages, which are compliant with the schema. This means that the function call is transmitted as an XML message. This also applies to the response and the error messages that may also need to be transmitted. Components that support or use Web Services can be implemented on any platform supporting XML and HTTP.

The technologies presented in this section (XML, SOAP and Web services) have not been defined by individual companies or groups of companies, but by the WWW Consortium (W3C). This is important particularly with regard to future OPC specifications. The unilateral orientation of OPC towards DCOM and thus towards Microsoft was often criticized in the past. The use of XML and Web services should eliminate this point of criticism.

Table 3.2 shows the states of the different specifications. "Recommendation" denotes a passed standard, "Draft" a standard still under discussion, and "Note" a working paper that is already relatively concrete.

Table 3.2: *XML specifications – Release states (February 2004)*

XML 1.0	W3C Recommendation
XML Schema Part 1 & 2 1.0	W3C Recommendation
SOAP/XMLP 1.2	W3C Recommendation
WSDL 2.0	W3C Draft

3.5 Implementation of OPC components

3.5.1 DCOM – Design concepts and possibilities of implementation

Using threads

The meaning of threads in the implementation of DCOM components was explained in section 3.2.1. What has to be observed in this regard when designing OPC applications?

Tables 3.3 and 3.4 show various possibilities of how functions can be allocated to threads. The tables evaluate these possibilities for client and server components.

These evaluations are not only important for your own implementations but should also be used for verifying components you are planning to buy.

Table 3.3: Allocation of functions to threads in OPC servers

Allocation of the functions to threads	Note
One thread per OPCGroup object	With many OPCGroup objects, this will lead to many threads but will leave the scheduling up to the DCOM runtime. This solution is not recommended.
One thread per OPC client	Only a few threads are created but the implementer is in charge of the synchronization. This solution is recommended. Race conditions may occur, however, if global resources are accessed simultaneously in such cases, access to the resource is blocked to other threads, which must wait.
One thread for all OPCGroup objects	Only one thread is created but the implementer is in charge of the synchronization. This solution is recommended if many clients (> 30) access the server. Race conditions may occur, however, if global resources are accessed simultaneously in such cases, access to the resource is blocked to other threads, which must wait.
Threads for data acquisition	In this case, care should be taken to ensure that the inter face between the server and periphery is optimally designed. Threads should only be used if acquiring the data takes much longer than managing the threads.

Table 3.4: Allocation of functions to threads in OPC clients

Allocation of the functions to threads	Note
One thread per OPC server	Only a few threads are created as long as there are few servers. This solution is recommended.
One thread for all OPC servers	Only one thread is created; the implementer is in charge of the internal management. This solution is recommended if a client accesses many servers (> 30). Race conditions may occur, however, if global resources are accessed simultaneously.
Thread(s) for operating the callbacks	These threads must execute few actions that take little time. No memory may be allocated. The creation of these threads depends on the application but cannot be influenced by the application.

Connection monitoring

What happens if the connection between a client and a server is broken by opening the network or switching off the PC?

Unfortunately, DCOM has a timeout of 3 x 2 minutes. This means that it takes 6 minutes for the component to receive the information that the connection is broken. This time period is unacceptable in automation technology. It is therefore essential to implement an appropriate connection monitoring solution in the client and server.

In the client, this is easy. A watchdog monitors whether calls from the server are received within a specific period of time. It can monitor asynchronous calls and also the use of callback methods. If the watchdog expires for the first case, you can immediately assume that the server is no longer available. If the watchdog for the last case expires you should check once again via a synchronous call (QueryInterface) whether the connection has been broken or whether there have simply been no changes in the process.

If the result of this call shows that the connection no longer exists, the client has to release the local resources.

The server can also monitor the connection. If the client has set up a bi-directional connection, the server can call the QueryInterface method for a non-supported interface. The corresponding return value must be received.

The OPC Data Access 3.0 and Alarms and Events 1.10 specifications now also define the possibilities a server has for monitoring the connection.

Implementation types for DCOM OPC servers

A DCOM OPC server can be implemented in three versions:

- DLL (InProc server): In this case, the server runs in the same process space as the OPC client. This is completely transparent for the client, as loading (and unloading) the DLL is performed by the DCOM runtime. Special peculiarities need to be observed for remote access to the DLL.
- Local or remote server (OutProc server): The OPC server is implemented as a standalone executable. It is launched by the OPC client. Many OPC servers today are implemented as OutProc server. The server can be located either on the same PC as the client or on a remote PC.
- Windows NT/2000/XP Service: A service is a special kind of OutProc server. It can be started immediately when booting the operating system and thus runs even without user logon. If the OPC server is implemented as a service, an existing user interface must be executed independently of the server.

Table 3.5 below shows the advantages and disadvantages of the individual implementation types.

Table 3.5: *Evaluation of the implementation types for DCOM servers*

	Advantage	**Disadvantage**
InProc Server	Best performance.	No simultaneous access by several clients if protected resources are used.
OutProc Server	Server can be connected directly with the configuration part.	Bad performance compared to InProc Servers.
Service	Runs even without user logon. Long-lasting initialization tasks (e.g. network configuration) can run in parallel to the boot process of the PC.	Not portable and cannot be used everywhere. The use of services is limited to Windows NT/ 2000. Section 3.6 deals with possibilities of developing OPC components also for non-Windows systems. For these systems, no service exists.

Web services are always implemented as executables. The specific manner in which they are implemented under a specific operating system is not defined.

What possibilities are there for implementing OPC components?

Figure 3.2 illustrates the memory layout of a DCOM server. Any language and development environment that allows the creation of such a structure can be used for creating OPC components. Examples are the languages C/C++ and Microsoft Visual C++ Compiler. You could use Borland Delphi and Borland C++ Builder also. Visual Basic is very well suited for developing clients that access the automation interface or use ActiveX Controls implementing client functionality.

3.5.2 Aspects of implementing Web services

In **Figure 3.8** you see the components that are required for implementing a Web service and a client that accesses it. To support coding and decoding the XML messages, an XML Parser is used. XML parsers are available for systems with practically unlimited resources as well as for systems with limited resources. The XML messages are exchanged using HTTP. Unlike DCOM, Web services can be implemented in almost any language. The only requirement, which is met by most parsers, is that the language has to support the integration of the basic functions (XML parser and HTTP).

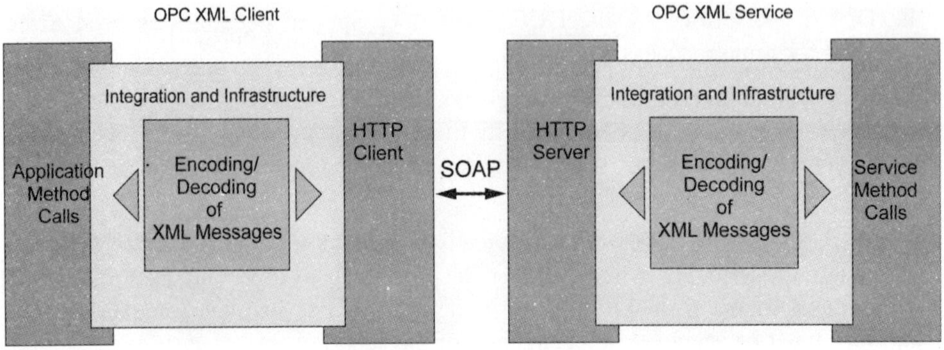

***Figure 3.8:** Elements of a Web service and a client*

OPC XML-DA can be implemented in three different ways:

- „DIY" implementation: HTTP and XML parser are available on the target platform or need to be ported there. Coding and decoding the XML messages, their exchange via HTTP, the definitions of the OPC specification, and the acquisition and processing of the data have to be implemented. This is very work intensive and requires in-depth knowledge.
- Web services are used in many areas, so that a large number of generally applicable toolkits are available. These toolkits allow you to a great extent to automatically generate the implementation for coding and decoding the XML messages and for exchanging them via HTTP from the WSDL file. What remains to be done is to implement the OPC specification and the manner of acquiring and representing the data. Please note, however, that these toolkits are not applicable to all environments.
- Use of OPC XML-DA toolkits. The toolkit implements both the general Web service aspects and the OPC XML-DA specification. All that needs to be done when using these toolkits is to implement the data acquisition and representation solutions. The toolkits can be ported and thus save you a lot of work.

There are no general recommendations for Web services with regard to the use of threads and the manner of implementation. The procedure depends on the specific target platform.

3.5.3 Creating OPC components by means of tools

Motivation

A simpler way to create OPC components is to use toolkits. Many parts of an OPC client and OPC server are common and can be generalized.

Toolkits encapsulate the DCOM, Web service and OPC functionality. They exchange information through an API with an application using the toolkits. Toolkits are available for client and server components. Most OPC DCOM toolkits support Windows 9X/ME/NT/2000/XP. Versions for Windows CE and Linux exist also. OPC XML toolkits should be portable to allow use on a wide range of different platforms.

The toolkits can be distinguished by their possibilities of optimization. There are compact solutions intended for a simple server type and more complex toolkits that permit extremely flexible solutions.

The following section explains the concept of one specific toolbox. (This toolbox is provided on the enclosed CD.) It contains both compact and complex solutions. For the complex solutions, the CD offers tutorials that quickly get you started. Programming samples are included for all versions.

The OPC Toolbox 4.00 comprises nine components which have common and specific parts. Four components are complete programs, which you can either easily integrate in your application (OPC Client Controls for *Data Access* and *Alarms and Events, Client Toolkit for .NET DA, XML-DA, HDA, AE, DX, .NET DA Server Toolkit*) or with which you can easily integrate your application (XPress OPC Server Tool).

The other components allow extensive optimizations and adjustments to various areas of application (dynamic namespace, memory, and performance optimizations). These four components are also referred to as the OPC C++ Toolkit. This toolkit forms a C++ class framework for creating OPC server and client applications. The toolkit supports OPC Data Access 1.0, 2.05, 3.0, OPC XML-DA 1.0 and OPC Alarms and Events 1.10.

The OPC C++ Toolkit contains general objects that correspond to the requirements of the specifications. There are the following common objects for the data access tools:

- Namespace (node, item, property)
- COM objects (server, group, item)
- I/O objects (cache, request, transaction)

The following common objects are provided for the alarms and events tools:

- Eventspace (type, category, attribute, ConditionClass, SubconditionClass)
- Areaspace (area, source, condition)
- Notification objects (event)
- COM objects (server, subscription, AreaBrowser)

From these objects, the toolkit then generates trees at runtime according to the specifications of the server or client application. The trees represent, for example, the namespace, the object hierarchy, or the eventarea.

What does the user have to do when creating different OPC components based on the C++ Toolkit?

Data Access Server and Alarms and Events Server

- Initialization of the toolkit.
- Creation of the object trees to be generated.
- Exchange of data with the process, in the case of Data Access Servers (reading and writing).
- Evaluation of the events (receiving data from the process and checking the event condition).

The libraries allow several clients to access the OPC server simultaneously. A separate thread is created for each client. All threads run in an MTA.

Other tasks, such as the cyclic update of values, filtering of events, browsing of the namespace, and much more, have been implemented in the toolkit. I/O tasks for reading the process variables can be optionally executed synchronously, asynchronously, or event triggered. The DCOM connection to the clients is monitored.

You can also create components that contain the functionality of the Data Access Specification, the Alarms and Events Specification, and OPC XML-DA all at the same time.

Data Access Client and Alarms and Events Client

The libraries allow an application to access several OPC servers simultaneously. A separate thread is created for each server. All threads run in an MTA. The threading model for the main thread can be selected as desired. User interface applications must use the STA in the main thread. Many clients will store the structure of a concrete application (including information on OPC servers and OPC objects). This is supported by the toolkit. Callbacks play an important part in the concept of client libraries. The callbacks are called by the toolkit when the corresponding events occur (e.g. data received from server, write method completed). The application needs to implement the corresponding functionality (e.g. output of the data on the screen, change of the background color of an input field). The toolkit performs all OPC communication tasks. In addition, monitoring and re-establishment of the connection is supported.

XPress OPC Server Tool

The XPress OPC Server Tool is a complete OPC server application. It is provided in two different variants – OutProc Server, and OutProc Server as service. The module needs to be extended by an application-specific DLL. The configuration of the server and the communication with the peripherals are tasks of this DLL. The DLL should export at least five and at most nine functions with predefined signatures.

OPC Client Controls for Data Access and Alarms and Events

The packages contain two ActiveX controls each. The *Data Access* package includes a control for data exchange (OPCDataControl) and a control for namespace browsing (OPCDATreeView).

The *Alarms and Events* package contains a control responsible for interaction with the server (OPCAEDataControl). Another control (OPCEventViewControl) can be used for the visualization of events.

The controls access the custom interface. They combine the simplicity of visual programming with the runtime efficiency of accessing custom interfaces.

.NET Client and Server Toolkits

The .NET client toolkit allows creating OPC clients in .NET environments. It supports access to OPC DCOM DA, HDA, AE, DX as well as OPC XML-DA servers. The .NET server toolkit allows creating OPC DA servers in .NET environments. Demo versions of these toolkits, together with a description and examples, are provided on the enclosed CD.

Figure 3.9: *.NET Toolkits for the development of OPC DA, HDA, AE, DX and OPC XML-DA products*

Table 3.6 below lists the development environments supported by the different toolbox components.

Table 3.6: *Supported development environments for the toolbox components*

	Visual C++ V6.0/.NET	**Borland Delphi V5.0**	**Borland C++ Builder V4.0**	**Visual Basic V6.0/.NET**
C++ Toolkits	C++		C++	
XPress OPC Server Tool	C/C++	X	C/C++	
OPC Client Controls	C++	X	C++	X

3.5.4 Creating OPC clients by using ActiveX Controls

ActiveX Controls are also DCOM objects. Among other features, they offer predefined interfaces over which they provide functions in a standardized manner. This makes these controls very easy to use.

As mentioned in section 3.5.3, the OPC Client Controls combine the simplicity of visual programming with the efficiency of accessing custom interfaces.

This section demonstrates by using two examples how easy it is to create OPC clients with the OPC Client Controls.

Visual Basic offers various data controls for displaying and writing values. The client controls can be connected directly with such a data control. As an alternative, the event implementation of an application can be invoked by the control.

The following steps need to be carried out to create an Data Access Client application with Visual Basic:

1. Define a form in the Visual Basic development environment.
2. Insert the OPCDataControl icon in the form.

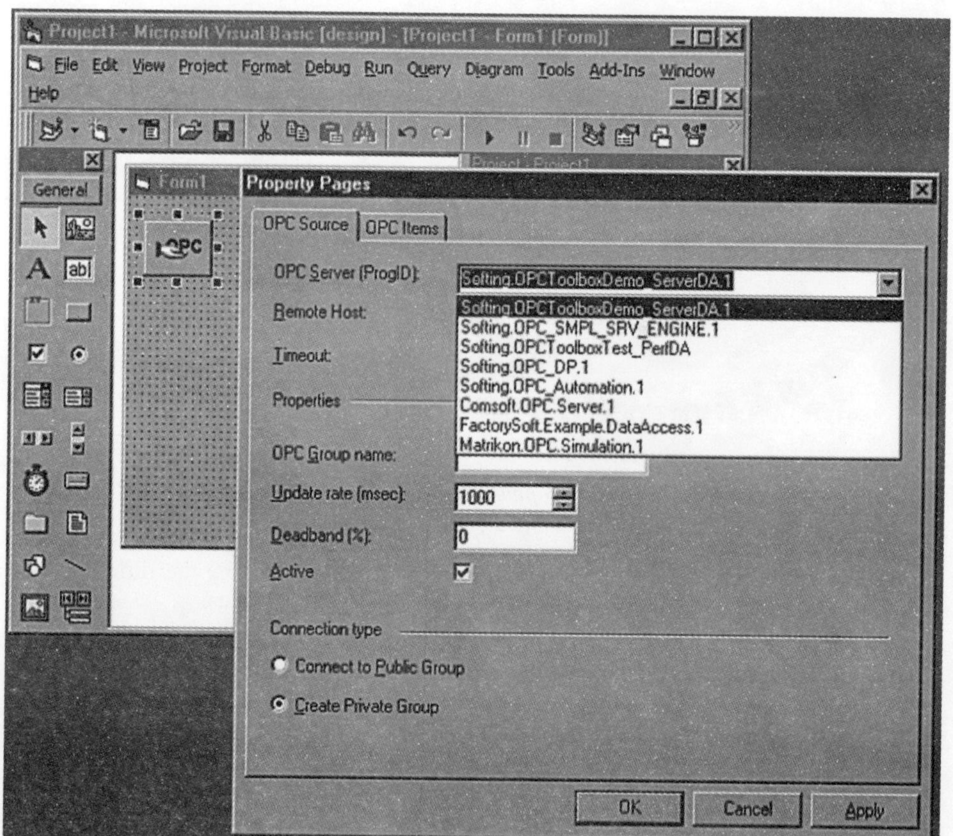

Figure 3.10: *Configuring a client application with ActiveX Controls*

3. Set the property values which determine the behavior and data sources for this client control (Name, ConnectType, ServerName, PercentDeadband, GroupName, Active, UpdateRate, TimeOut). The result of these steps is shown in **Figure 3.10**.
4. Implement the Load method for the form. In the implementation, Connect needs to be called at the DataControl. This launches the corresponding *Data Access Server*.
5. Create text boxes in the form. The contents of the DataField property correspond to the fully qualified ItemID.
6. To transmit the entered values to the server, the application needs to call the control's Update method. This requires adding a command button to the form and implementing the Click method for this element.
7. Labels can be used for displaying data. Here again you need to set the DataField property. The corresponding fully qualified ItemId must be located there. The result of these last steps is shown in **Figure 3.11**.

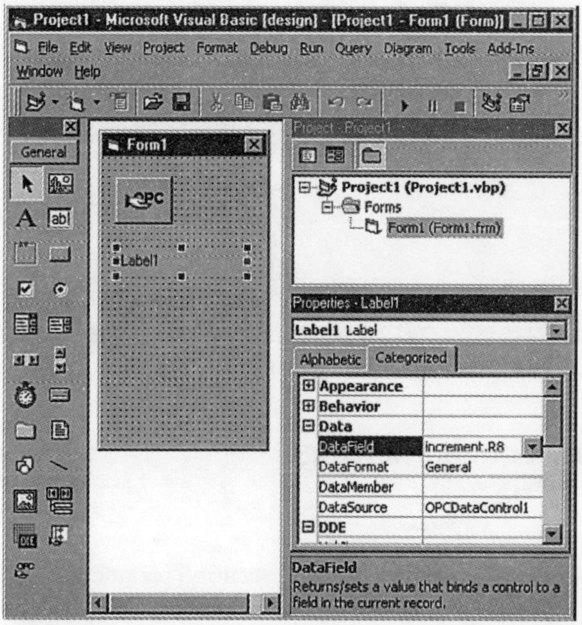

Figure 3.11: Defining the properties of the client's display element

To implement an Alarms and Events Client application with Visual Basic, perform the following steps:

1. Define a form in the Visual Basic development environment.
2. Insert the OPCAEDataControl and the OPCEventViewControl in the form.
3. Set the property values for the OPCAEDataControl, which define the server and the event sources. The result of these steps is shown in **Figure 3.12**.
4. Set the property values for the OPCEventViewControl. These settings determine how the events are displayed.

5. Invoke the Connect method at the OPCAEDataControl, e.g. within a Command_Click method.

Creating OPC clients can be that easy.

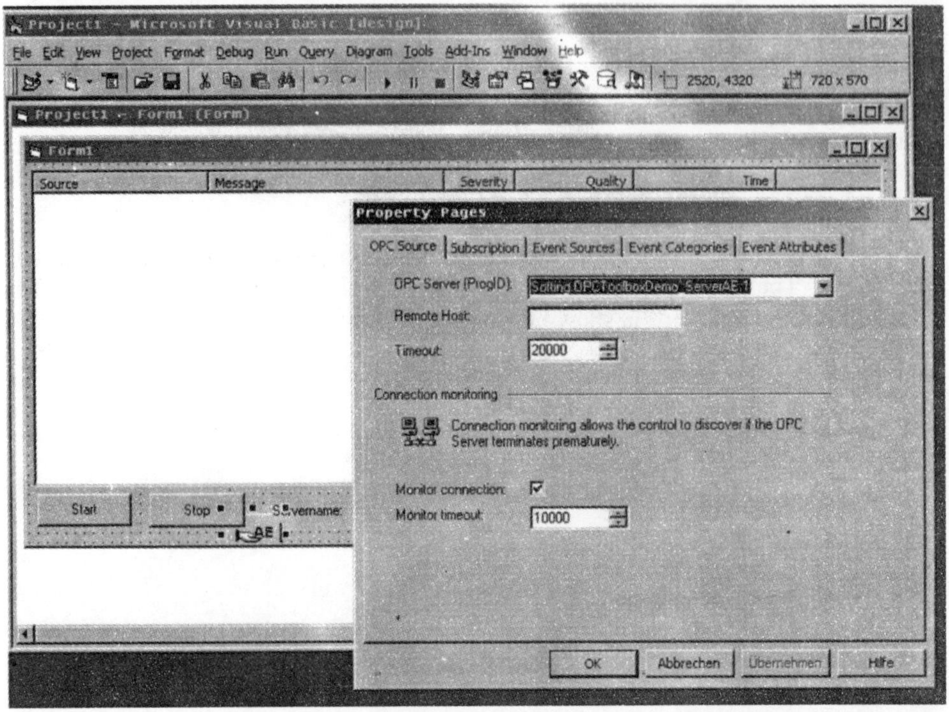

Figure 3.12: *Configuring an Alarms and Events Client with an ActiveX Control*

3.6 Implementation of security settings

3.6.1 DCOM security

This section discusses the possibilities of setting various security parameters and describes their effects.

Security parameters can be set in different ways:

- By using the DCOMCNFG program. To use this program, the user must have Administrator rights. This ability to define security settings is referred to as declarative security. All settings apply either to the system or to a complete DCOM component (InProc or OutProc Server). It is not possible to define settings for parts of the components (e.g. single objects, interface, methods).
- By using DCOM Security API methods. Settings can be made by the client or the server or by both. Security settings implemented by means of these methods cannot be overwritten from the outside (via DCOMCNFG). This procedure permits a finer granularity of security objects. Settings can be defined even for very small parts of the component (method call, data source, data sink).

Declarative security

First the functionality of DCOMCNFG and the consequences of different settings will be explained. **Figure 3.13** shows the startup page of DCOMCNFG. The information offered by the program and the settings made by the user are stored below the component's ApplicationID key in the registry, as described in section 2.2.2 (see also **Figure 2.1**).

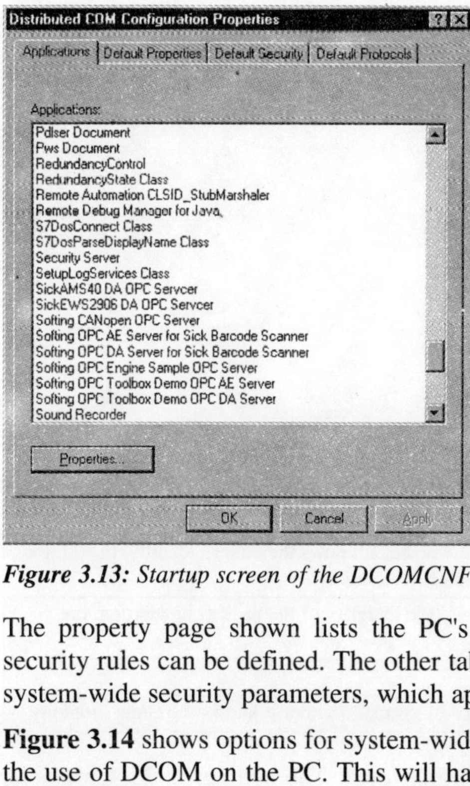

Figure 3.13: Startup screen of the DCOMCNFG program

The property page shown lists the PC's (D)COM components for which special security rules can be defined. The other tabs and dialog boxes allow you to influence system-wide security parameters, which apply to all (D)COM components.

Figure 3.14 shows options for system-wide parameter settings. Here, you can disable the use of DCOM on the PC. This will happen very rarely. The entry "Enable COM Internet Services on this computer" should be enabled in order that DCOM can also run over HTTP.

This is not the same as the Simple Object Access Protocol (SOAP, s. paragraph 3.4) but a proprietary solution by Microsoft.

The level on which the client is authenticated by the server and vice versa is set in the Default Authentication Level list box. There are several levels, each of which enhances the strictness of the authentication check. **Table 3.7** lists the individual levels and explains their meanings.

Section 2.12.1 introduced the NT Security model. It showed three different possibilities for securing the communication path between active instances (client and server). These possibilities are reflected in the different authentication levels.

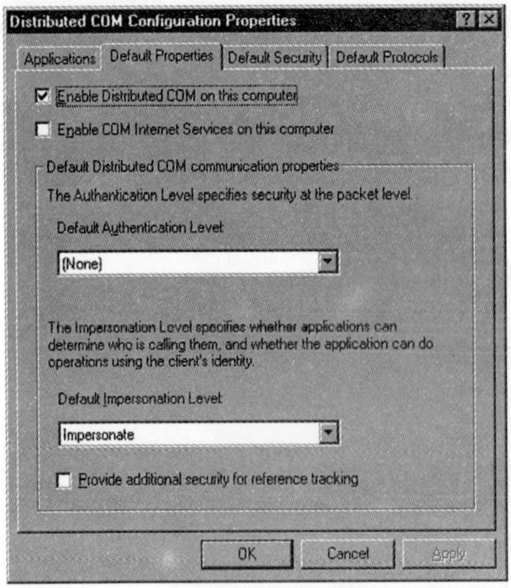

Figure 3.14: *Default Properties tab*

Table 3.7: *Authentication levels in DCOM*

None	The server does not check whether it knows the client's identity.
Call	The server checks for every method call whether it knows the client's identity.
Packet	The server checks for every data packet whether it knows the client's identity.
Packet integrity	The server checks for every data packet whether it knows the client's identity. In addition, it checks whether the integrity of the packet is assured.
Packet privacy	The server checks the client's identity and the integrity of the packet. In addition, the packet is encrypted.
Default	Corresponds to "Connect."
Connect	Each time a connection is set up, the server checks whether it knows the client's identity.

In this tab, you can also set the impersonation level. The setting of this parameter specifies the degree to which the server can use the client's identity when it needs to perform operations at the client's request. Examples showing the necessity of setting the impersonation level are:

– The server is to access resources available only for the client's user, e.g. files.

– The server has been implemented as a service and runs under the system account. In this case, a corresponding impersonation level is required to invoke methods at the client. A process that runs under the system account can only use local resources.

– The server must access identity information about the calling client. This is related to the authentication level settings.

The impersonation level is always assigned by the client and therefore needs to be set on the PC on which the client is running.

Table 3.8 lists the possible levels and explains their meanings.

***Table 3.8:** Impersonation levels in DCOM*

Anonymous	The server cannot determine the client's identification information.
Identify	The server can read the identification information and execute several functions.
Impersonate	The server can use the client's security context.
Delegate	The server can invoke calls to other servers at the client's request.

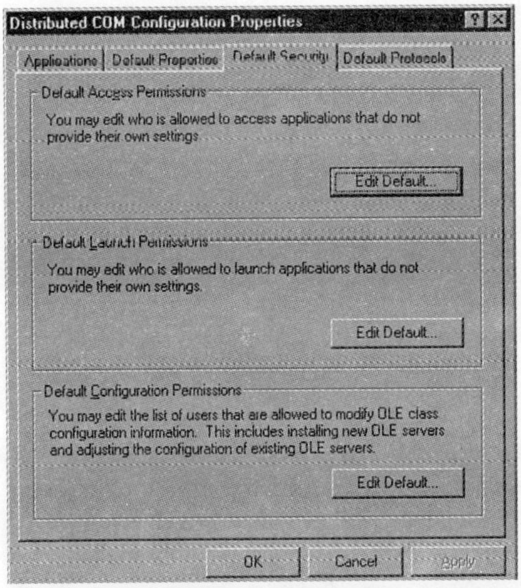

***Figure 3.15:** Default Security tab*

Like the previous tabs, the tab shown in **Figure 3.15** is used to define security settings for all DCOM components. You can determine which processes, depending on their user identifications, are allowed to

– launch servers (launch rights)

– access servers that have already been launched (access rights)

Client users can be defined either within a domain or a workgroup. User administration in a domain is easier and more flexible, as the definition of users is done only once in the global domain controller.

The tab settings shown in **Figure 3.16** again apply to all DCOM objects. The user can set the transport protocols available on the PC. The user can also set the order in which they will be used to set up a connection with a DCOM object on another PC (i.e. the order in which they are listed). When should these settings be considered? Different transport protocols can be supported on different PCs. The DCOM runtime on the client PC will try the protocols in a set order when connecting to the server PC. If the sets of protocols differ, e.g. if a protocol on one PC is not supported on the other PC, it can take some time for connection information to be passed at runtime. It can be that the time taken to connect becomes greater then the time available for launching a server. In this case the DCOM runtime will stop attempting to launch the server.

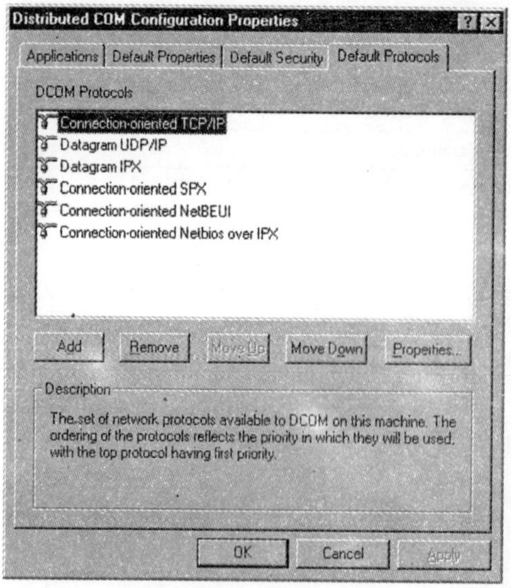

Figure 3.16: Default Protocols tab

Transport protocols use ports to address applications. If there is a firewall between computers running clients and servers, normally only specific ports will be opened. You can use the tab settings shown in **Figure 3.17** to determine that the ports opened by the firewall match the ports via which the DCOM components interact.

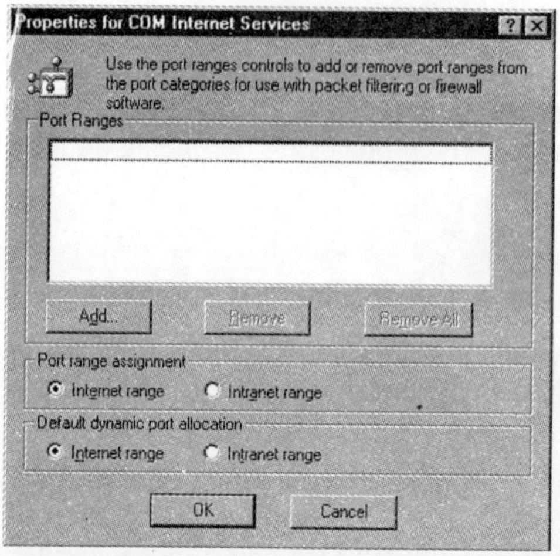

Figure 3.17: COM Internet Services Properties tab

The tabs shown in the following figures allow you to set the security properties that apply specifically to one component. First, you can redefine the authentication level for the component, if desired. This allows you to fine-tune the authentication level for individual components. The tab can be seen in **Figure 3.18**.

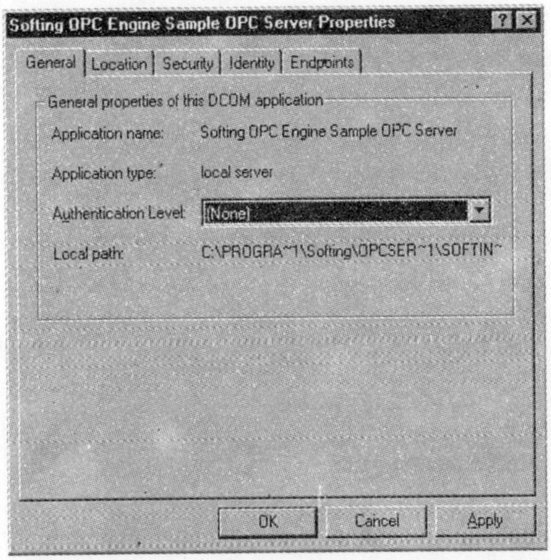

Figure 3.18: General tab for a DCOM component

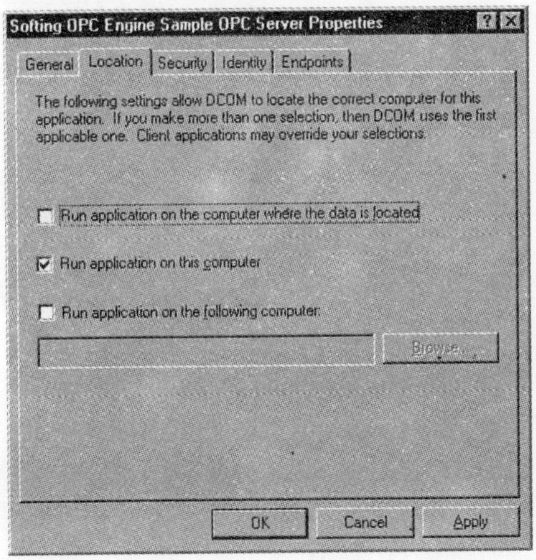

Figure 3.19: Location tab for a DCOM component

Figure 3.19 shows the Location tab, with settings that have a rather historical meaning. DCOM had not yet been implemented in the first version of Windows 95, which meant the client could not indicate on which PC the server was to be launched. By activating the corresponding checkboxes in this tab, you could set the PC on which the server is located. This had three drawbacks:

1. Selecting the PC could not be performed at runtime.
2. Entries for the server that are not necessary, such as the CLSID, had to be added to the registry.
3. Clients could only interact with a single server for this CLSID (exactly one server on a specific PC).

Most DCOM components will require the user to set special access and launch rights for them. The corresponding tab can be seen in **Figure 3.20**. To determine the users that are permitted to launch or access a component, you can use the tab shown in **Figure 3.21**.

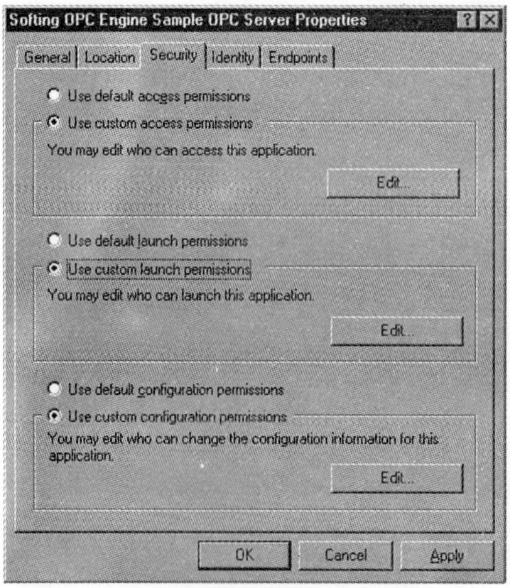

Figure 3.20: Security tab for a DCOM component

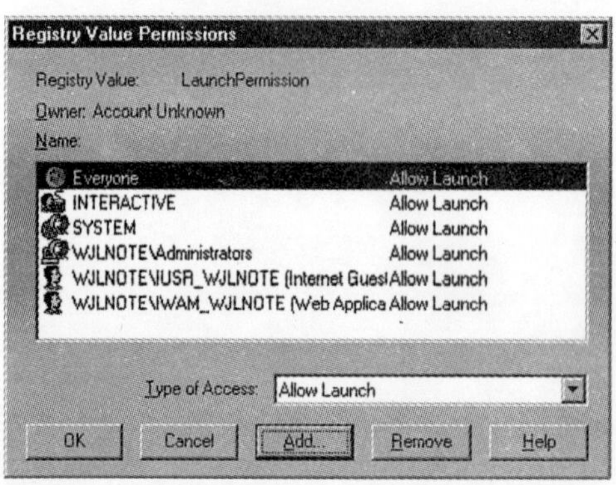

Figure 3.21: Tab setting access rights for a DCOM component

After setting the access and launch rights, you can now define the account under which the component is to be run. **Table 3.9** lists the four possibilities, explains and evaluates them.

Table 3.9: *User accounts for the DCOM component*

Setting	Meaning
Interactive User	The user currently working at the PC. If this value is set and a client starts the component from a different PC, the user working at the PC will see a user interface, if there is one. This is not desirable in most cases.
Launching User	Here, the server runs in its own environment. The user working at the PC will not see a user interface if there is one. If many clients launch the server, resource problems may arise, as a new environment is created for each client. In each environment, a new server is launched. Each of these environments is a new process. For this reason, only the first server launched is granted access to exclusive resources.
This User	Here, any user can be entered. If a client launches the server, an environment is created for the corresponding user. All other servers run in the same environment, i.e. in the same process. On the one hand, there are no access problems; on the other hand, a user interface, which may exist, cannot be seen. This setting should be applied.
The System Account	This setting is only available for DCOM servers that have been implemented as a service.

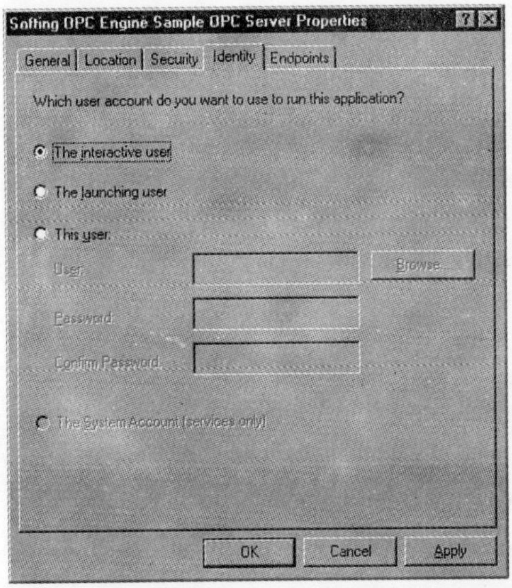

Figure 3.22: *Tab setting user identity for a DCOM component*

Figure 3.22 shows the tab for setting the user identity.

Apart from determining the ports for the transport protocols for the entire system, you can also do this specifically for an individual DCOM component. The tab can be seen in **Figure 3.23**.

The Microsoft operating systems differ in the available Dcomcnfg settings and in the way the program is started. In Windows95/98/ME, DCOM is not fully implemented and, therefore, the remote launch of servers is not possible. Consequently, there are no authentication level settings for servers. Users rights for launching the servers cannot be assigned either. The server has to be launched before the client can access it.

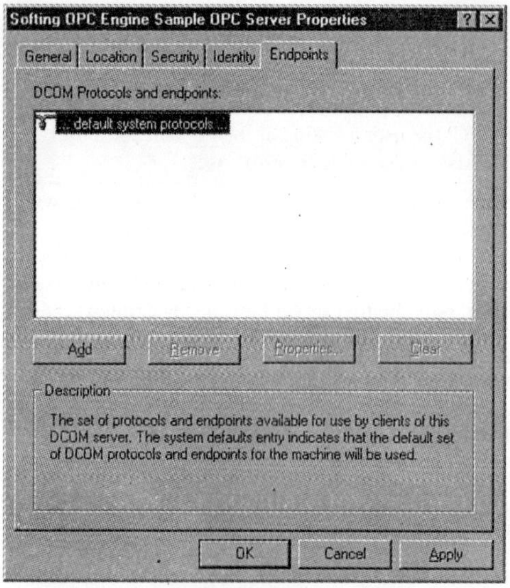

Figure 3.23: *Tab for setting the endpoints of the transport protocols*

Under Windows NT/2000/XP, the described functionality is fully available. Under Windows XP, Dcomcnfg is called via Control Panel entries. With SP 2 for Windows XP, a very restrictive security policy is installed in the PC. A white paper describing the necessary settings for running OPC applications is available for download on the web site of the OPC Foundation (status: November 2004)

The description of DCOMCNFG has now been completed. Before proceeding with notes on programmatic security, a few scenarios for defining the security settings with DCOMCNFG will be explained.

Several scenarios for launching and accessing DCOM servers have been examined, concerning different user accounts and impersonation and authorization level settings. The aim was to:

– find settings that allow launching and accessing of components

– detect correlations between problems that occur during access and launch and these settings

Table 3.10 explains the different scenarios.

The first scenarios investigate, if clients are allowed to launch and access servers. The clients run under different user accounts.

Table 3.10: *Scenarios for defining security settings with DCOMCNFG*

Conditions	Result	Explanation
Server and client are in the same domain. The same user has logged on to both PCs. Different authorization and impersonation levels are tried out. The server has been implemented as an OutProc server. The server runs under the identity of Interactive User.	The client can always launch the server and read and write data. This means that callbacks are set up also.	The users are in the same domain and the server runs in Interactive User mode. With regard to security, this scenario is identical to a local access.
The server is in a domain; the client is in a workgroup. The client's user has logged on to the workgroup PC with an account that is known in the server's domain.	The client can always launch the server and read and write data. This means that callbacks are set up also.	Server and client are not in the same domain but the client runs under an account that is known in the domain. This case is thus a variation of the above scenario.
The server is in a domain; the client is in a workgroup. The client's user has logged on to the workgroup PC with a local account not known in the server's domain. The different authorization and impersonation levels are tried out.	The client can launch the server and read and write data only when the authentication level is set to "None." With all other authentication levels, the client cannot receive any data from the server.	The server checks the client's authenticity before it returns the call. Since the server does not know the client account, the check yields an incorrect result. The server therefore will not return any calls to the client. The authentication level is only checked after the server has been launched; the server is therefore launched.
The server is in a domain; the client is in a workgroup. The client's user has logged on to the workgroup PC with a local account not known in the server's domain. The server runs under the identity of This User. This identity is identical to the workgroup's user. The different authorization and impersonation levels are tried out.	The client can launch the server and read and write data. This means that callbacks are set up also.	By setting This User and making the user known, the server knows this identity and can authenticate the client. This corresponds to the case of an external user who wants to access a DCOM server in a different domain. The positive result, however, presupposes that the external user has logged on under an account that is known in the server's PC.
Server and client are in the same domain. The user has logged on to each PC with a different account. The different authorization and impersonation levels are tried out. The server has been implemented as an OutProc server. The server runs under the identity of Interactive User.	The client can always launch the server and read and write data. This means that callbacks are set up also.	The users are in the same domain and the server runs in Interactive User mode. The difference between this and the first scenario lies in the different accounts.

Conditions	Result	Explanation
The server and client are in a not-trusted domain below another domain. A different domain user has logged on to each PC. The different authentication and impersonation levels are tried out. The server has been implemented as an OutProc server. The server runs under the identity of Inter active User.	The client can always launch the server and read and write data. This means that callbacks are set up also.	If two not-trusted domains belong to the same domain, the exchange of user identities is performed over this domain. The user accounts are thus known in both domains.
The server and client are in two different domains. The server runs under the identity of This User. This user is only known on the server PC and is not the same as the client user.	The client can only launch the server and read and write data if the authorization level has been set to "None." With all other authorization levels, the client cannot receive any data from the server.	Here again the server detects a conflict between the identity under which the server is to run and the identity of the client user.

The last scenario corresponds to server access over the Internet. The server is located in a domain, And for external access it runs under the identity This User. A specific user is defined. In the case of external access, the client can launch the server. The server will not call back the client and thus will not set up any callbacks as the identity of This User will conflict with the account of the user who launched the client. The client is thus restricted to synchronous calls unless the authentication level in the server is set to "None." This client-user constellation and these settings in the server occur whenever a new user tries to launch the server. In applications on the Internet, this is often the case.

Programmatic security

As well as setting security aspects with DCOMCNFG, you can do so with DCOM Security API functionality.

Two types of function can be distinguished:

- Functions called when components are started and that define security settings.
- Functions called while the component is executed and that can query and define security-related settings.

The former class of function will be described first.

CoInitializeSecurity

This method can be called once for each process. This distinguishes it from CoInitalize(Ex) which was explained in section 3.2.1 and which needs to be called for each thread. If security settings are to be preset in the thread or for a method call, this must be done with one of the methods described further below. If this method is not called by the component, the DCOM runtime uses the default settings entered in the registry under the corresponding AppId. **Table 3.11** lists the parameters of CoInitializeSecurity and their meanings. CoInitializeSecurity can be called by the server and by the client.

Table 3.11: *CoInitializeSecurity parameters*

Parameter	Meaning
SecurityDescriptor	This parameter assigns the access rights of the component. This can be performed in different ways. The parameter can receive either a reference to a Win32 security descriptor, or a reference to an AppId under which the security has stored information, or a pointer to the IAccessControl interface at an object.
AuthenticationService	This parameter specifies the authentication service to be used. With Windows NT, only one parameter can be set here. Windows 2000 supports two possibilities.
AuthenticationLevel	This parameter contains the authentication level which the client would like to use and which is the minimum level required by the server. The higher of the two values will then be used during runtime.
ImpersonationLevel	This parameter is only used in the client since it defines the extent to which the server is allowed to use the client's rights.
AdditionalCapabilities	Here, additional security settings can be defined. The parameter value can specify, for example, how the contents of the first parameter (SecurityDescriptor) are to be used.

COAUTHINFO structure

This structure is part of the COSERVERINFO structure, which is a parameter of the CoCreateInstance and CoGetClassObject methods, among others. Obviously, this structure is thus only used in the client implementation.

Table 3.12 lists the structure elements and describes their meanings.

Table 3.12: *Elements of the COAUTHINFO structure*

Parameter	Meaning
AuthenticationService	This parameter specifies the authentication service to be used. With Windows NT, only one parameter can be set here. Windows 2000 supports two possibilities.
AuthorizationService	This parameter defines the authorization service to be used.
ServerPrincipalName	Name of the server to be used in connection with the authentication service.
AuthenticationLevel	This parameter contains the authentication level the client would like to use.
ImpersonationLevel	This parameter is only used in the client as it defines the extent to which the server is allowed to use the client's rights.
AdditionalCapabilities	Currently not used.
ClientIdentity	This parameter passes a user identity (logon, password, domain name) to DCOM.

Whereas CoInitializeEx defines security settings for the entire component (client or server), the elements of this structure allow definition of the security related to a client-server connection.

The remaining part of this section will explain the methods used to receive or set security-related information during server or client execution.

Table 3.13 shows the methods available in the server. The server can use the functions only within a method called by the client.

Table 3.13: Available security API methods in the DCOM server

Method	Meaning
CoQueryClientBlanket	With this method, the server can query the client's security credentials. This method would certainly be applied in implementing the OPC Security Specification.
CoImpersonateClient	The server temporarily uses the client's security credentials.
CoRevertToSelf	Resets the security credentials to those that the server had before CoImpersonateClient was called.

Table 3.14 explains the methods that are available in the client.

Table 3.14: Available security API methods in the DCOM client

Method	Meaning
CoQueryProxyBlanket	The client queries the current security settings.
CoSetProxyBlanket	The client modifies the current security settings.for a specific connection to a DCOM object in a different process. The client could change the authentication level.
CoCopyProxy	This method creates a copy of a proxy, which is not influenced by, for example, CoSetProxyBlanket.

The pseudo code given in **Table 3.15** shows how to use the above methods.

Table 3.15: Use of declarative security in the DCOM client and DCOM server

Client	Server	Note
CoInitializeEx	CoInitializeEx	Initialization of DCOM.
CoInitializeSecurity	CoInitializeSecurity	Definition of the basic security settings.
CoCreateInstance		Definition of the security settings that are effective for this connection to a server.
CoSetProxyBlanket		Setting of security specific for a call.
	CoQueryClientBlanket	Query of the security settings in the call.
	CoImpersonateClient	Server impersonates client.
	CoRevertToSelf	Reset of the security settings in the call of CoImpersonateClient.
CoSetProxyBlanket		Reset to the previous settings.

The description of the DCOM security aspects has now been completed.

3.6.2 Security in the context of Web services

The use of DCOM as the technological basis for OPC specifications has limitations on the implementation and use of the products. The previous section may have given you the impression that security is a quite difficult issue with DCOM. This impression has been confirmed again and again by practical experience over the years. In the following section, you will see that the use of Web services makes things much easier in this regard, as well.

Possible security solutions for Web services

1. Use of basic authentication

Basic authentication is a procedure that you will know from accessing web sites. Before a page is displayed, you are asked to authenticate yourself by entering your user name and password. The server can then check which rights this user has. Basic authentication is part of HTTP and does not require additional implementation. The transmitted data need to be evaluated on the server side. A drawback is that the user name and password are transmitted in plain text, which is hardly acceptable. Therefore, the functionality described next is used to eliminate this drawback.

2. Encryption of the authentication data

Besides HTTP, there is HTTPS, which is the secure variant. Here, data are transmitted in an encrypted way. This also applies to the authentication information. Using this solution requires some implementation work in particular environments. OpenSSL provides a portable library in the public domain, which provides assistance in this regard.

Besides attempts to gain unauthorized access to data, there are also attempts to prevent the use of functionality or to affect the computer in other ways. The solution described next provides assistance in this regard.

3. Use of firewalls

Firewalls can be used to limit access to the computer and to prevent attacks.

Web services are internationally standardized. This standardization also includes the specification of a Web service architecture. Web service security is an integral part of this architecture and is described next.

4. Web service security

The specification of this aspect has not yet been completed. The solution shown in the **figure 3.24** would be a possible approach.

As described above, Web service security is based on open, generally accessible solutions that ensure a logical and comprehensible behavior.

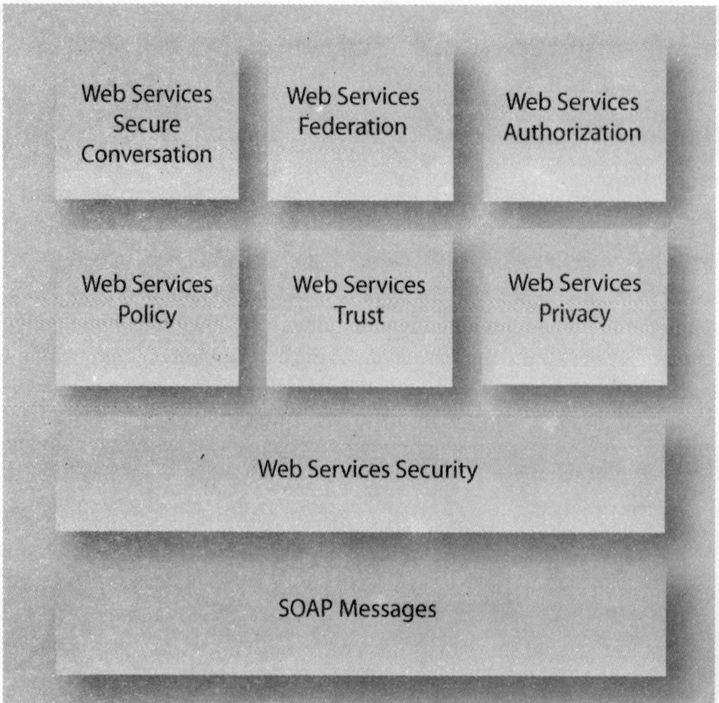

Figure 3.24: *Possible security approach with Web service security*

3.7 OPC in Embedded Systems

PCs using the Windows 9X/NT/2000/XP operating systems are currently the main area of application for OPC components. Automation technology also offers a variety of other platforms, however, on which programs are running that provide or access data. There was hardly any possibility so far to integrate these platforms into OPC applications.

This has changed decisively with the availability of OPC XML-DA.

3.8 Summary

This chapter is definitely the most demanding of the book.

You might want to read the first section several times. Nevertheless, a few questions may remain. You may have come to feel that DCOM can hardly be understood, let even be programmed.

This feeling may change, though, once you have read the subsequent sections. Then you might think: Actually, programming DCOM is logical and clear-cut.

Both are true – DCOM is not easy to understand, and still it is possible to create programs. Gaining the know-how and experience takes much time, however.

This know-how and experience are contained in the toolkits. In using these to create OPC products, two worlds meet: those of the OPC specialists and the users of this know-how, who know their world of devices and visualization programs very intimately. The use of toolkits greatly facilitates the creation of OPC components and requires less study. Clients can also be created very easily using special ActiveX Controls, as described in section 3.5.4.

The use of Web services as an additional technological basis for the OPC specifications will greatly expand the choice of possible target platforms for clients and servers. It is difficult to say how difficult implementation will be in the individual cases. One thing is for sure, however: Portable toolkits are an interesting and efficient alternative.

When the components are finished, they should eventually be put to use. They will not always be run on the same PC, though. This will invariably bring up the issue of security concepts. Even though the main focus in the beginning might have been to access the server with the client, the task of a targeted use of security concepts will come up with time. The contents of section 3.6 are intended to assist you in solving both tasks.

4 Application

4.1 Introduction

All areas of automation technology today are marked by growing requirements on availability, productivity, and quality. With productivity and quality rising, availability generally needs to increase and operating and maintenance costs need to decrease.

On examination, the importance of specific requirements differs for different areas of industry. The following sections will take a close look at the requirements of manufacturing industry, process industry, and building automation.

A characteristic feature of the manufacturing industry is the production of lot sizes and/ or the execution of recurring work steps which are often independent of each other. The degree of decentralization within a production plant today is increasing. Traditional manufacturing systems consist of a supervisory and control system, which distributes the production parameters to programmable logic controls (PLC), which in turn communicate with digital or analog subassemblies and other intelligent field devices.

On average, the number of data points to be controlled or measured ranges around several thousand.

The requirements placed on the communication between PLC and peripherals are very high. Many areas, such as robotics or the measuring and testing industry, demand stringent real-time requirements with cycle times of less than 10 milliseconds. The area of drives often requires an equidistant data transfer, for example, for axis interpolation.

Even though a three-shift operation is quite common in the manufacturing industry and machinery thus runs around-the-clock, the requirements with regard to fail-safety are only average. In many cases, stopping the production if a device fails is less expensive than making the entire system fully redundantly and running a hot-standby[13] strategy. In this case it is important, however, to ensure that the maintenance personnel can replace the defective components within a few minutes and then resume production. Higher requirements on fail-safety apply to areas in which dangers for persons can arise, for example, with combustion controls, press controls, or turning lathes.

In the food processing, chemical, and pharmaceutical industries, and in the process industry and process technology in general, the production is mainly batch-oriented. These batch processes are continuous.

---

13 Hot-Stand-By is the ability of an application to switch automatically to a redundant device when the default device fails without interrupting the controlled process.

Typically, very complex plants are predominant in the process industry. Implemented as distributed systems, they are highly decentralized and extend over wide areas. The data quantities that occur in such plants can reach up to several hundred thousand data points. The requirements on central supervisory and control systems are thus very high as far as the handling of the data quantities is concerned. The aspect of time, in contrast, is less critical and often lies in the seconds range.

A fundamentally important criterion in the process industry is high availability. Due to the complex and lengthy start-up procedures of the continuous processes, which can often take many hours, the plants are not shut down. Expensive redundancy concepts with hot standby prevent error-related process interruptions. The requirements in the area of maintenance and commissioning are correspondingly high. Changing over or extending the plant must be possible during the running operation. Additional security requirements arise in explosion-hazard surroundings, such as in petrochemistry, transport, and the industry of gases.

The requirements in the heating, air-conditioning, and ventilation technologies – in fact, in energy and building management in general – constitute a third category. As in the process industry, very great data quantities also occur in building automation. When several building complexes, with hundreds of rooms and thousands of light switches, sensors, and temperature control devices, are monitored centrally, the data quantities often reach many hundred thousand data points.

The aspect of time can usually be neglected: it is not really decisive whether information that the room temperature has been exceeded, for example, is reported in five or ten seconds. The decisive requirements in building automation lie in the handling of the data quantities in the case of an alarm. If there is a fire, several sensors will transmit messages simultaneously, which in turn will result in a flood of signals for triggering the sprinkler system, informing the fire station, automatically closing the windows, and other actions. The outcome is a "burst" of data which must not be lost and for which the chronological order in which they arrive must not be changed. In a few areas, such as in medicine or clean-room technology, very high demands are additionally placed on security with regard to the failure of devices, for example, in the operation room.

These quite different criteria demanded for the various industries justify the existence of several very different OPC specifications. *Data Access Clients* and *Servers* were the first OPC products. Their functionality and performance are very well suited for use in the manufacturing industry. In this industry, OPC is by far the most popular. A clear delimitation still needs to be defined for the use of OPC in areas with stringent real-time requirements and equidistance, for example, in robotics or drive control. These areas require operating systems with hard real-time capabilities and DCOM support as well as an OPC interface that has been optimized for real-time requirements, as specified in the *Data Access 3.0* workgroup.

The *Historical Data Access Specification* and *Batch Specification* meet the special requirements of continuous and batch-oriented processes.

In building automation, the advantages of the *Alarms and Events Specification* come in useful. This area, however, also uses HDA servers for backup and recording.

In the following sections of this chapter, several OPC Client and Server products will be described. Some of the products dealt with can be used as an OPC Client and OPC Server at the same time. In each section, the products or product family will be introduced with their areas of application and intended use. Where OPC is applied will also be shown. The articles have been submitted by employees from other companies. Each author will be named at the end of the corresponding article. The assessments of the advantages and disadvantages of using OPC have been made by the individual authors and need to be viewed against the background of the given OPC implementation.

In an additional section, several OPC end users will describe their experiences in using OPC. These articles illustrate the wide range of possible OPC applications.

The chapter concludes with a discussion of further aspects of use. One section presents the results of performance tests. The last section is devoted to the possibilities of using OPC on the Internet.

4.2 OPC Server Product Samples

4.2.1 CANopen OPC Server – Easy Integration of Fieldbus Systems

A large number of fieldbuses have become indispensable in the automation world today. The advantage of fieldbuses is that devices from different manufacturers can communicate via a standardized protocol. Their scope of application reaches from simple process data exchange between field devices in production plants to drive controls with equidistant data transmission and deterministic time requirements.

All fieldbuses share specifications describing the communication properties and parameters of the field devices.

During fieldbus commissioning, two types of settings must be carried through:

- Configuration: setting of bus parameters, which specify the communications protocol and describe the physical properties of the network.
- Parameterization: setting of the device parameters as they are defined in the device descriptions and can be adapted depending on the application.

Access by PC application software to the devices within a fieldbus network, e.g. for visualization or configuration, takes place by means of a PC plug-in card with a protocol driver specific to the operating system. This protocol driver is activated via an API (Application Program Interface). The protocol drivers with API differ depending on the fieldbus and the manufacturer.

A prominent feature of OPC is the concealment of these manufacturer- and fieldbus-specific properties.

Using the CANopen protocol and a CANopen OPC Server as an example, the characteristics of a fieldbus protocol and the corresponding mapping to the OPC Server interface are explained below:

Figure 4.1: Access to CANopen devices with a CANopen PCcard

CANopen is on the march

The Controller Area Network (CAN), formerly employed only in the automotive industry, is today extensively used in measurement technology and control engineering. With its application protocol layer and device profiles, CANopen is today widely distributed in building automation, traffic and railroad engineering, robotics, building of special machines, medical engineering, studio technology, and other areas. Many leading drive manufacturers use CANopen as their system bus.

The performance characteristics comprise:

- Multi-master capability
- Excellent real-time quality
- High interference immunity
- Good possibilities of extension
- A very favorable price/performance ratio

With CANopen, any station within the network may send data to any number of other stations at any point in time. Information for several destination addresses only passes the bus once. Furthermore, there is event-driven notification of the individual network stations' applications. These modern procedures are highly efficient, reduce the workload on the bus and on the respective applications, and thus lead to excellent response times.

In CANopen, data are mapped to individual frames as variable objects. In terms of data transmission, the CANopen specification distinguishes between process data communication, which must take place in cycles of very high rates, and parameterization or service data communication. Correspondingly, process data are transferred in

PDO (process data object) and configuration data in SDO (service data object) frames. In addition to PDO and SDO communication, there are a number of network management services, e.g. for the starting and stopping of devices and for transferring the device status.

Electronic Data Sheets (EDS) define device-specific properties and indicate which PDO or SDO objects are supported by a device.

From an EDS file, the individual data objects – with their ID numbers, data types, etc. – can be defined depending on the application. These specifications are stored in a Device Configuration File (DCF).

By the mapping to OPCItem objects via an OPC Server interface, the specific characteristics of frames and their data objects can be concealed from the application programs (with OPC Client interfaces).

Figure 4.2 shows how process data objects of a CANopen device are mapped to OPCItem objects.

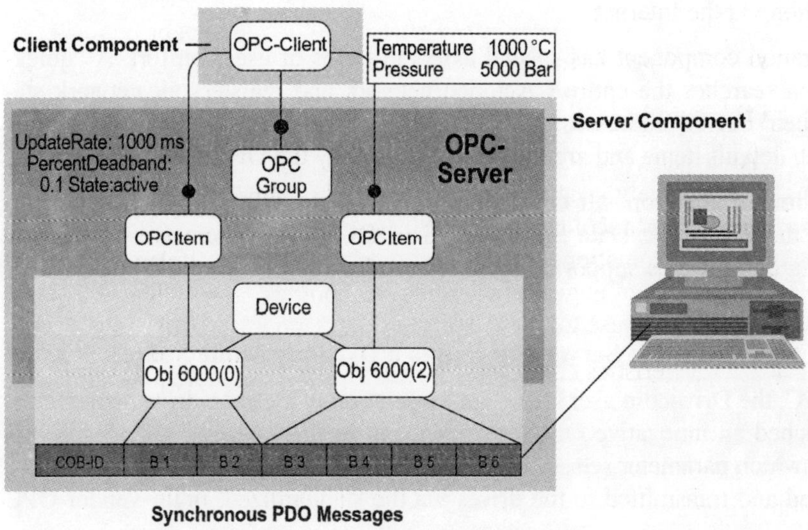

Figure 4.2: Mapping of PDOs to OPCItem objects in the CANopen OPC Server

In an OPC Server, service data objects and network management data can be mapped to OPCItem objects in the same manner and can be made available to any OPC applications via the standardized OPC interface.

The CANopen OPC Server as a Modular System

The CANopen OPC Server can be started as a service which runs on Windows NT. With this service, runtime and GUI are decoupled; therefore, GUI-relevant functionalities were swapped out to separate components.

The CANopen OPC Server was developed on the basis of the OPC Server Development Kit by Softing and therefore supports all OPC interfaces in accordance with the *Data Access Specification* version 2.04. It functions as a CANopen Master according to the CANopen Specification DS301 and supports PDO frames for cyclic data exchange and SDO frames for parameterization. A testing mode for the commissioning phase permits reading and writing of SDO frames. During the commissioning phase, variable objects can be assigned any default starting values. This is essential in many plants.

A diagnostic and monitoring component provides the user interface for diagnostic information and process data. For this purpose, diagnostic and process data are exported to a file within configurable intervals. Transmission of the diagnostic information and process data via a DCOM interface is in preparation.

The diagnostic and monitoring component offers a comprehensive display of the current data. In this way, it is possible to check the status of the entire network, of the individual devices, and of process data points during commissioning.

As data is described in the XML format, it can be displayed with any WWW browser supporting XML. This results in different possibilities of OPC Server remote diagnosis. The operating states of the OPC Server and of the devices, for example, are sent to a control station via the Internet.

The configuration component has special assets in terms of user comfort. A "quick-start" function searches the entire CANopen network and registers all network stations with their device addresses. The registered devices are represented in tree topology with default items and are thus ready for use by the OPC Server.

Through an import function, all configuration data and device parameters are imported from the Electronic Data Sheets (EDS or DCF file). There are only a few manual configuration steps supported by drag-and-drop and by copy-and-paste functions.

Mapping of the OPC Drive Server Concept to a Dynamic Name Space

Based on OPC, the Drivecom user group, an association of German drive manufacturers, has launched an innovative OPC drive concept. In this concept, a namespace is defined with which parameter sets, or drive programs with vendor-specific drive tools, can be created and transmitted to the drives via the standardized, multi-vendor OPC interface.

An OPC Drive Server forms a layer between the different OPC Servers for fieldbus systems and an application program with OPC Client interface. This intermediate layer contains functionality specific to a lower-layer Fieldbus OPC Server as well as functionality specific to the respective drive manufacturer, e.g. for the configuration or for starting and stopping drives.

Thus, the OPC Drive Server covers all requirements specific to the company and the communication application. The user, who can access the OPC Drive Server with any kind of OPC Client application, can exchange drives and the communications interface without any undesired side effects. OPC Client applications can read process data, download parameter sets to drives, and start and stop programs. For this purpose, the OPC Drive Server creates OPCItem objects in the Fieldbus OPC Server. In defined semantics, the OPC Drive Server transmits information concerning the devices and communication properties to the Fieldbus OPC Server. In this manner, a namespace dependent on the respective application is created dynamically in the Fieldbus OPC Server.

The CANopen OPC Server supports this "OPC profile" for drive manufacturers. Mapping of parameter sets to SDO frames takes place by means of identifiers, which are transferred to the CANopen OPC Server via the OPCItem semantics.

Standardized Access to Drives

The Lenze GmbH & Co KG Company uses the CANopen OPC Server with integrated OPC drive profile. The Lenze drives, as well as field devices by other vendors, are connected to a supervisory and control system via the company's own CANopen system bus. For configuration and parameterization of the drives, a parameterization tool is employed. It is used to create drive programs and parameter sets. By the OPC interface of the CANopen OPC Server, the programs and parameter sets are loaded into the drives. The name space of the CANopen OPC Server is created according to the semantics of the OPC drive profile. Drives can be started and stopped via OPC. Process data, status, and parameters can be read and written by any OPC Clients via the OPC items of the dynamically created OPC namespace.

As is demonstrated clearly by the example of the CANopen OPC Server, properties specific to the fieldbus and the manufacturer are concealed behind the OPC interface.

Fieldbus OPC Servers can be equipped with additional intelligence for automatic recognition of all devices in the network and for carrying through settings for communication with the field devices by means of diagnostic frames and by reading in Device Description Files. In addition, comfortable project-specific settings can be performed via the OPC interface by means of dynamic namespace configuration.

This is the component-based, intelligent concept of the Softing Fieldbus OPC Servers, which cannot only be applied to CANopen OPC Servers but also to PROFIBUS, DeviceNet, and other fieldbuses.

Thus, configuration, commissioning, and maintenance are greatly simplified and the costs reduced accordingly.

Jürgen Lange, Softing AG, Munich, Germany

4.2.2 LonWorks® OPC Server – Connecting Two Open Standards

What is LON?

LON – Local **O**perating **N**etwork – also known as LonWorks®, was developed in 1990 by the American company Echelon. The protocol is defined according to all seven levels of the ISO/OSI model, and is standardised as ANSI/EIA/CEA 709. It runs primarily on what are known as the Neuron® chips. A LON network, which is also known as a "control network", operates on the multi-master principle, in which the devices exchange their messages on the basis of events. Up to 32,385 devices, the "LON nodes", can participate in one logical network.

In addition to twisted pair wiring (Twisted Pair, Free Topology, Link Power), the power supply network, radio, infra-red and optical fibre can all be used as the LON transmission medium.

Where is LON used?

There are already a million implementations of the LonWorks® technology in various parts of the world. It is used, for instance, in cable television networks, on ships and railways, in traffic routing technology and for many other purposes. In addition to its industrial applications, however, the main area where LON technology is applied is in building automation, particularly for medium and large functional buildings. Nevertheless, LON is being increasingly used in domestic applications, for example in electrical devices such as washing machines and refrigerators.

How did LON come about?

As mentioned above, LonWorks® technology was developed by the firm of Echelon in 1990. Mike Markkula, who is also one of Apple's founders, went on to found Echelon, with the aim of creating a consistent basis and an open technology through which very different devices could communicate.

The history of the development of LON technology is marked by a number of important milestones, including

- the development of the Neuron® chip, which includes the LonTalk protocol as a primary element in its firmware,
- the founding of LonMark® International to propagate the technology and to act as the body responsible for ensuring interoperability, and
- the development of the LNS™ architecture (LonWorks® Network Services) on Microsoft Windows© operating systems.

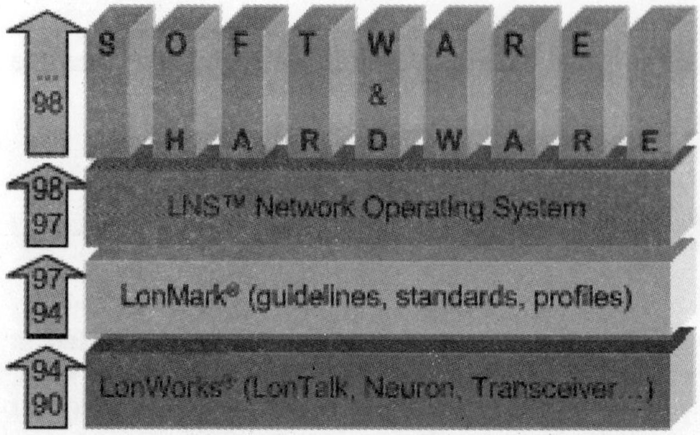

***Figure 4.3:** LonWorks® milestones*

Each of the illustrated layers represents a specific system generation. The lowest step quickly led to a large number of applications, most of which, however, had proprietary properties. Only the Neuron® chip and the LonTalk protocol were used, in order to hold down the company's own development costs. In the second step, LonMark® International brought interoperability to products from different manufacturers. LonMark® International has a similar significance for LonWorks® technology as does the OPC Foundation for OPC technology.

The problem that until then had remained unsolved, however, lay with the methods and procedures (generally referred to as tools) with which networks could be set up, extended and serviced. Life is relatively easy for competitors from other network technologies who only specify a single tool. As a result of the wide range of applications on which LonWorks® had set its sights, however, it was clear from the outset that a single tool, or even tool set, would never be enough. This is what led the company's thinking in the direction of a tool basis, that is of a uniform network operating system. This was the birth of LNS™.

What is LNS™?

LNS™ – LonWorks Network Services – provides the foundation for interoperable tools and modular applications for LonWorks® networks. For the first time, it introduced the performance features of a client-server architecture and of an open software architecture based on software components for control networks. This gave rise to a new generation of co-operating tools and applications, with which LonWorks® networks can be installed, serviced, monitored and controlled.

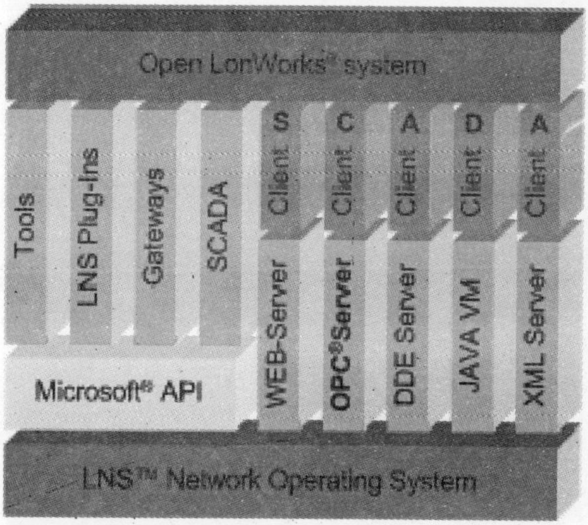

Figure 4.4: *LNS™ as a network operating system*

The advantages of the client-server architecture have made this into the standard for data networks. Almost every data network used within companies (intranets) today employs a client-server architecture. The internet is, without a doubt, the largest client-server system in the world today. LNS™ brings the benefits and advantages of the client-server architecture right to the world of control systems, and can even use these as an element in its own control structure. LNS™ is the client-server network operating system for automation, regardless of whether a building or an industrial system is involved. The many restrictions of conventional bus systems or hierarchical networks are in this way overcome.

LNS™ is built on a compact, object-oriented programming model that significantly reduces development time, as well as reducing the demand for memory resources and processing power in the host computer. LNS™ represents the network as a hierarchy of objects that provide a set of services, have a number of properties, and communicate changes through events.

At the time of writing this article, the latest version of the software is “LNS™ Version 3.0”, but the next generation, bearing the name “LNS™ Turbo Edition”, is already waiting in the wings.

Why an OPC server?

The development of an OPC server for LonWorks®, with LNS™ as its basis, is the logical consequence. On the one hand, a LNS™ OPC server is a client to the LNS™ server, and on the other hand it is the server for the OPC client. In this way it provides an elegant link between the two open standards, thereby opening the LON world to hundreds of OPC-enabled standard products such as engineering displays, office applications and control systems.

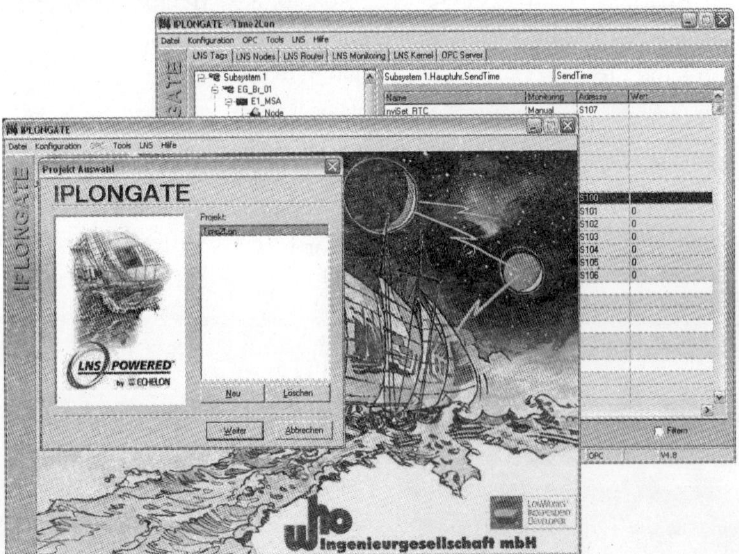

Figure 4.5: The IPLONGATE LNS™ OPC server

IPLONGATE is an OPC server of this type based on LNS™. It was originally developed by the French company of IP Systemes in the year 2000. IPLONGATE was first used in one of the world's largest LonWorks® networks at the Coeur défense in Paris. The network, at that time still administered under LNS™ 2.0, comprises about 23,000 LON devices.

Since 2003 IPLONGATE has been continued and further developed by the German company of "who Ingenieurgesellschaft mbH". "who" is one of about 20 companies in various parts of the world who have been granted the status of "LonWorks® Independent Developer" (LID) by Echelon. It carried out the porting of the OPC server from LNS™ 2.0 ot LNS™ 3.0, and gave the server many new performance features. As a member of the OPC Foundation, "who" also carried out certification of the server.

IPLONGATE is one of the first servers that has been tested in accordance with the Data Access 3.0 specification from the OPC Foundation. IPLONGATE received the "Compliance Tested" logo as early as 2003. The fast, straightforward implementation of the OPC specification was made possible by using the OPC Tool box from Softing.

As an LNS™ client, IPLONGATE uses the full functionality of the properties, methods and events of the LNS™ architecture. This gives the OPC server access to all the data items such as network variables and configuration properties in the LON network. The OPC items are represented in the OPC area in accordance with the LNS™ hierarchy.

Depending on the medium used, reading data from the LON network is an important task for the server. Cyclic interrogation of a large number of data points would heavily load the available bandwidth, particularly over slower media such as Powerline, with its 4.5 kbps, or Free Topology, with 78 kbps. The operation of IPLONGATE therefore, like LonWorks® itself, is oriented around system events. In other words the LNS™ server informs the OPC server about the change in a data point, and IPLONGATE correspondingly makes this available.

Since the great majority of a data points in LON systems are implemented as what are known as SNVTs (Standard Network Variable Types), and since these are usually structured, IPLONGATE also permits the individual structure elements of these data points to be accessed. When we bear in mind that, for instance, a network variable may well consist of 64 individual elements, and that it is possible for each device to have 10 of these variables, and that, moreover, there may be 100 devices in the network (and that a network comprising 100 LON devices is in reality on the small side) we can see that a large figure – in this case, 64,000 OPC items – is soon reached. Particular attention was and is given to this in the further development of IPLONGATE.

On top of this, additional information about the LON network, such as the status of devices or of routers in the OPC area, is made available.

In the course of further development for the next LNS™ generation, the Softing OPC Toolbox is being used to give IPLONGATE an interface to OPC XML DA 1.0. The next version of IPLONGATE will not just be a simple modification, but will be a generation change that will fit the LonWorks® OPC server for the latest technologies, so that it can continue to link the two open standards in the future.

Dipl.-Ing. (FH) Andreas Banz, who Ingenieurgesellschaft mbH, Lübeck, Germany

4.2.3 Softing 4CONTROL – PC-based control with OPC

Softing AG has been active in the area of industrial automation for over 20 years. Valued as an innovative and highly qualified development partner, Softing has developed automation systems from engineering via control through to visualization and integration for a variety of leading enterprises in automation technology. Starting in 1999, Softing has established the product family 4CONTROL as a flexible, modern and highly appreciated automation platform for different industry segments.

The product family 4CONTROL offers different controller types. All of them are supporting the OPC standard with an OPC server. Among these controllers, the so-called "embedded controller" (no moving parts like fan, HDD) are developed for different applications with a high degree of decentralisation. Softing supplies with the engineering system based on the programming standards of the IEC61131-3 and the runtime system with the corresponding OPC server also the base for customized or custom specific controllers.

The 4CONTROL PC based Control has a special role, since OPC client as well as OPC server functionality is integrated.

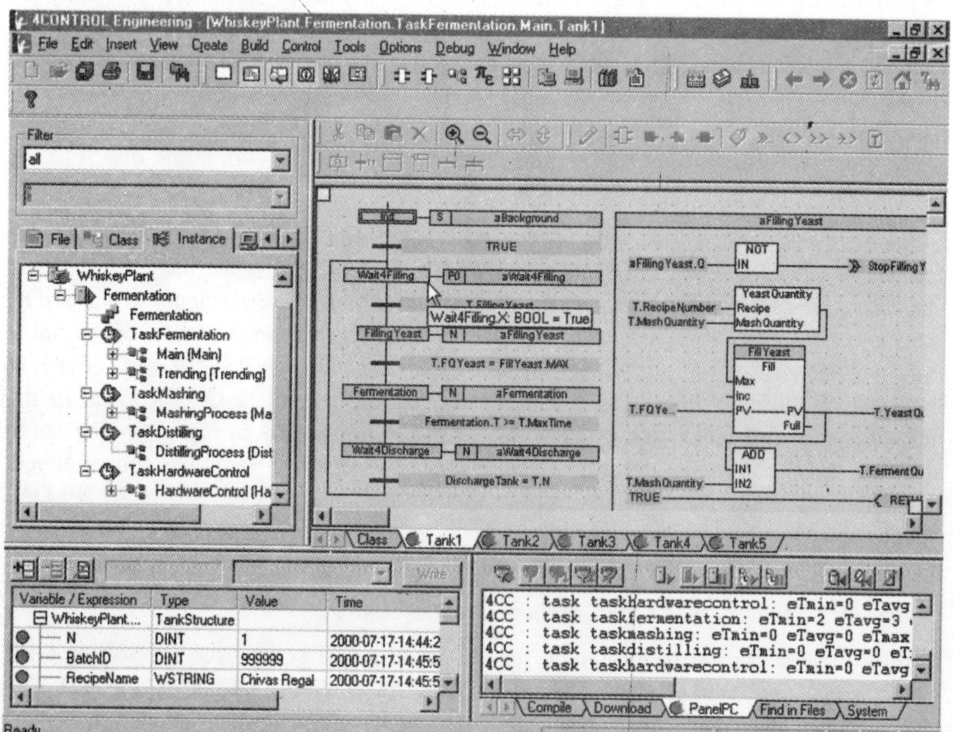

Figure 4.6: 4CONTROL offers an IEC 61131-3 engineering platform for distributed control concepts

4CONTROL PC based control is a PC technology based automation system that offers a complete solution platform from controlling, programming, and integrating through to visualizing. A further product range represents 4CONTROL Embedded, a programming and control system for programmable automation components. With its programming system to IEC-61131-3, its web browser-based visualization, and, not least, with its OPC Client and OPC Server interface, 4CONTROL consistently supports open standards.

The openness of this solution on the base of a PC with a Windows (e.g. XP) environment allows the implementation of applications with high standing requirements for communication and data processing. It is therefore possible to collect data from several OPC servers, to perform the control and data processing and to deliver via the integrated OPC client the generated information to other systems (OPC gateway functionality).

This enables all users to implement just the type of control system they need without having to consider manufacturer-specific restrictions. Key technologies such as COM, DCOM, and ActiveX are consequently just as readily available as traditional Windows services such as I/O management, access protection, virtual memory, or cache.

Distributed control – Process control in the network

Distributed control, combined with the integration of an OPC periphery, ideally emphasizes the advantages of a software control like 4CONTROL: It allows the distribution of control and visualization tasks to several networked PCs, as desired, and the extension of existing plant controls without involving great technical efforts. By adding further PCs or 4CONTROL embedded controls during the running operation, functions or I/O connections can be extended. Plant alterations and maintenance work can be carried out without having to shut down the entire plant.

Further, the distribution over several PCs or 4CONTROL embedded controls allows the control of large and small plants alike (from 100 up to several 10.000 physical data points). The hardware that is used for the specific control task can be cost-efficiently adjusted at any time. In addition, the hardware for the individual parts of the plant can be implemented with exactly the performance necessary to meet the specific requirements. This is an advantage that, for example, makes free resources available for the use of other applications on the control PCs, like visualisation, MES or simple databases for archiving historical data.

Running the control tasks parallel on several PCs, offers the possibility of redundancy via OPC. The diagnostic functions also become more flexible: A field engineer with the necessary access rights can log on to PC-controlled plants and call diagnostic and maintenance pages over the Internet. With minimum system requirements, the status of the plant can thus be influenced from practically anywhere. For this purpose, 4CONTROL goes for cost-efficient standard solutions. In contrast to proprietary solutions, Windows provides all the necessary basic services right from the start, thus making the distributed control functionality possible.

Users profit from the enormous increase in the performance of PC hardware – their investments in the software are saved and the staff does not constantly need to be newly trained. The connection between the internationally standardized Internet technology and the open OPC interface gives users back the freedom to employ existing and new technologies totally flexibly and independently of specific manufacturers.

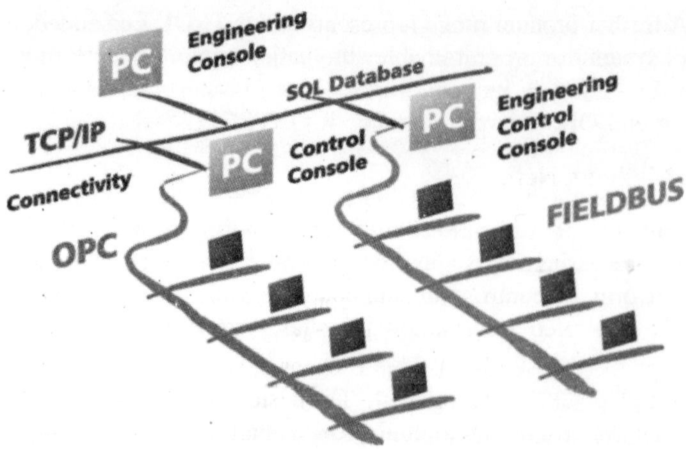

Figure 4.7: Distributed control architectures with 4CONTROL

OPC Server integrated

The "zero engineering" visualization feature, which is already included in the 4CONTROL system, is automatically generated when compiling the application programs (IEC control programs) and can be represented by any standard Internet browser from any desired location. Wherever this built-in visualization is not sufficient, users will want to use a visualization system package from a SCADA manufacturer. The OPC interface, which has by now been implemented in nearly all SCADA systems, allows the application-specific visualization of the 4CONTROL control as an OPC Server via these packages. All variables, function blocks, programs, and tasks contained in the IEC control program can be represented in the visualization as OPC items; user entries are transferred directly to the control program via the integrated 4CONTROL OPC Server.

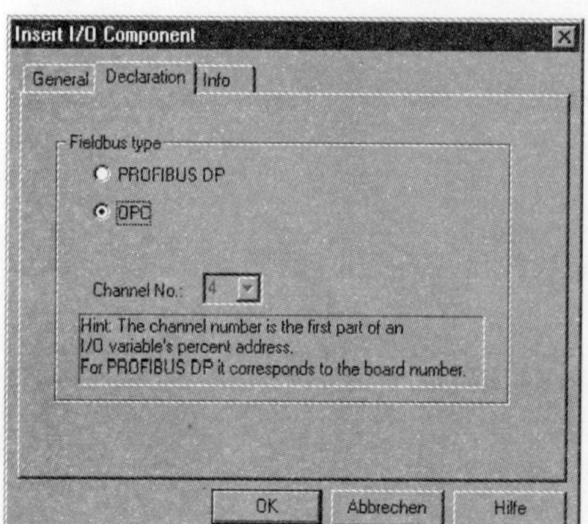

Figure 4.8: OPC connections are implemented in just a few steps with the help of 4CONTROL wizards

With the new standards like OPC and 4CONTROL as the control program, the hardware for the individual parts of a plant can be implemented with exactly the performance necessary to meet the specific requirements.

A concrete example: Integrated distribution and invoicing system of the Bavaria Brewery, Lieshout, Netherlands

With an annual output of 4.5 hectoliters of beer and 1.5 hectoliters of soft drinks, the Bavaria Brewery is the second largest brewery in the Netherlands.

The brewery's beer distribution to restaurants has been converted from barrel delivery to tanks. To optimize the scheduling, each tank truck has been equipped with an operator panel based on an industrial PC (IPC). This IPC has been provided with the PC-based control 4CONTROL as the core component. The customer database and the faceplates created with Visual Basic have additionally been installed.

Figure 4.9: *Control unit and drawing equipment on the Bavaria tank truck*

4CONTROL takes the customer and delivery data directly from the brewery's SAP-R/3 system, controls the calibrated drawing equipment via PROFIBUS and a direct RS232 connection, takes over the driver's entries on the operator panel, balances the customer database, and operates the invoice printer. OPC plays an important part as the link between the most diverse applications, which here, on a common platform, perform the overall task of the integrated distribution and invoicing system for the Bavaria Brewery. Thanks to OPC, no proprietary adjustments among the individual components were necessary in this application, which considerably cut the project planning and handling costs.

Oliver Roth, Softing AG, Munich, Germany

4.2.4 RAPID™ – Historical and Batch Data Acquisition with OPC

About Automsoft

Automsoft, founded in 1997, is a specialist database company, which develops and markets industry-specific plant management products to process manufacturing companies. These products include the core of the product family, RAPID as well as RAPID-Pharma, RAPID-Food, RAPID-Oil and Gas with a number of others in development. Its customers include many of the top 20 Pharmaceutical companies in the world, as well as a number of global Food, Chemical, Oil and Utility companies. It has offices in USA, UK, Ireland as well as distributors covering mainland Europe, USA and Asia-Pacific.

RAPID

Traditionally, Plant Information Management Systems (PIMS) or Plant Historians (Historians) have fallen into three categories:

1. Proprietary historian products that interfaced to a DCS, SCADA or PLC system through a specially written driver.
2. Add-on historian products from the major DCS and SCADA vendors, which worked only with their systems.
3. Historical log files produced by SCADA/HMI systems, which were stored in a proprietary fashion: these log files had special tools that extracted data to a spreadsheet, before analysis could be carried out.

None of the above satisfies the requirements of the today's business environment. This new environment requires that data be available immediately and easily to its users' computers. By effectively analysing historical data, businesses can plan for the future and learn from past mistakes. RAPID is a time-and-batch solution that answers the need of this new economy for a fast, powerful and easy to use PIMS.

Proper and reliable storage of historical process data is fundamental in the operation of pharmaceutical, food and beverage companies. Consumers must know that manufacturers have adhered to the strictest quality standards in the process. If an error enters the process, manufacturers must be able to pinpoint and solve the problem quickly to prevent widespread disaster.

Since the arrival of new technologies in this sector – namely high performance, complex database systems – the PIMS is seen as the link between plant floor systems and ERP systems, and in some cases pushing data and information directly to the Internet.

This has come about due to change in focus among technology users: mainly a desire by the customer for more open and interoperable systems.

PIMS products have always captured all the activities in a manufacturing plant, but they never had the open interfaces to freely exchange this information with applications in other parts of the plant. However, thanks to industry standards, such as OPC Data Access and OPC Historical Data Access, this is now possible with a new generation of software products such as RAPID.

RAPID Architecture

RAPID is a state of the art Plant Information Management System that offers the highest level of performance in the marketplace today. RAPID is used in the process manufacturing and utility industries. It provides manufacturers with the ability to view online the status of the production process, in addition to the complete historical record of the manufactured product. Through its batching functionality, RAPID has the ability to trace through production process data and identify each of the elements of the process, batch by batch.

The architecture of RAPID is outlined in **Figure 4.10** below. RAPID is an n-tier component based software suite built on XML, COM and OPC technology with a fault tolerant object oriented database management system (OODBMS) as the historical data repository.

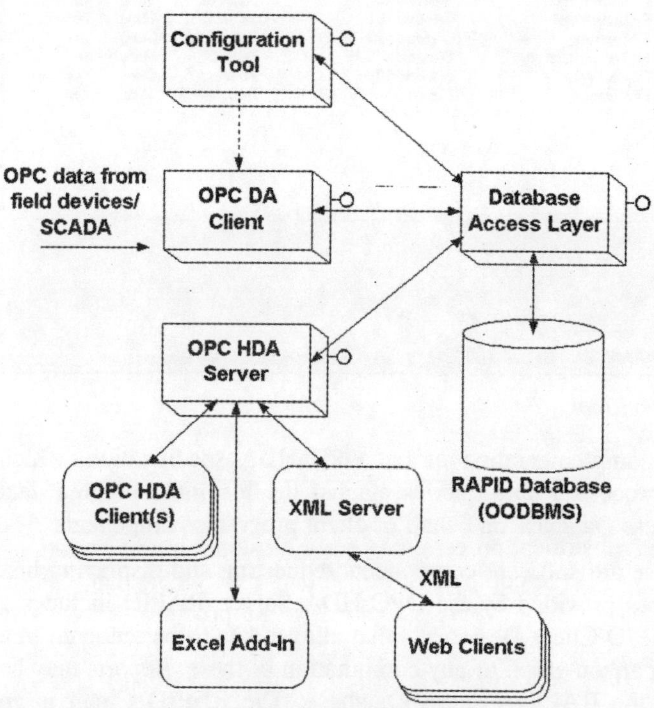

Figure 4.10: Detailed RAPID Architecture

The function of each (major) component is as follows:

Data Abstraction Layer (DAL): Provides database services on behalf of the other RAPID components e.g. insert and retrieve events for an OPC Item.

OPC DA/A&E Client: Collects the process data and alarms and events from OPC DA and A&E Servers e.g. from a SCADA system, DCS or a PLC, and stores this data via the DAL in the RAPID Process Database.

Configuration Tool: Allows the user to configure which items and OPC DA/A&E Servers to collect process data from. Note that this meta-data is also stored in the

RAPID Process Database. In addition, configuration changes may be made online i.e. without the need to stop collecting data. **Figure 4.11** below shows a sample screen shot.

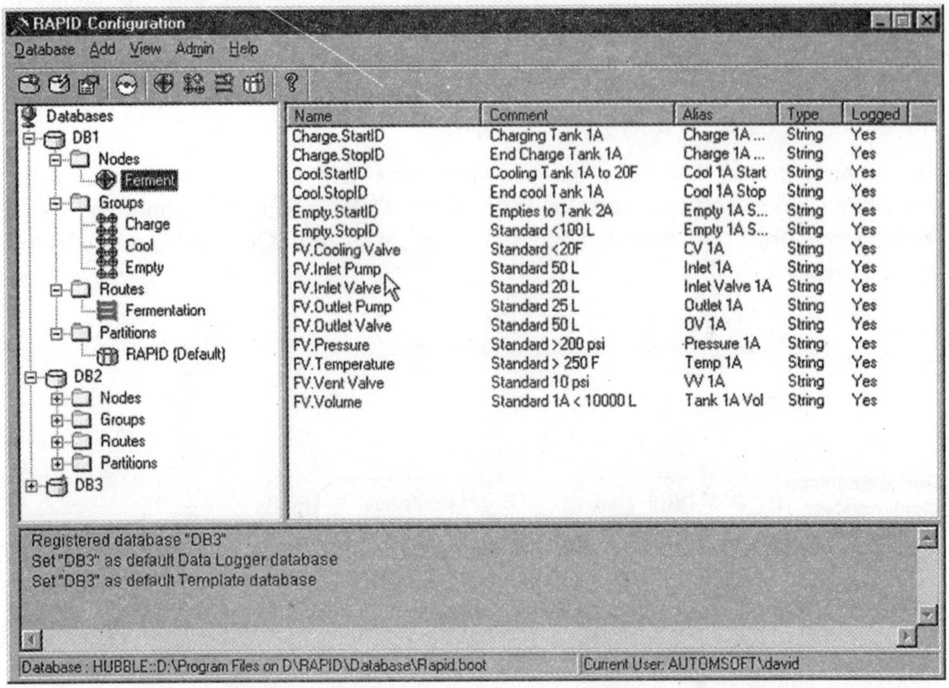

Figure 4.11: The Configuration Tool

OPC HDA Server: Is an implementation of the OPC HDA specification, which performs both raw and processed data queries against the historical process data stored in the RAPID Process Database on behalf of client programs/components.

OPC HDA Client(s): These are software components requesting and displaying historical data via the services provided by the OPC HDA Server. RAPID includes a powerful client tool – RAPID Client Designer – that allows data to be rendered in a spreadsheet, a trend/ comparison graph or any combination of these. Reports may be saved in PDF format in the RAPID Process Database. The RAPID Client is an ActiveX component and so may be embedded into any ActiveX container e.g. many leading SCADA systems. **Figure 4.12** below shows a sample batch comparison report created with the RAPID Client. Of course, third party OPC HDA clients may also be used.

Excel Add-in: Is an 'add-in' for Microsoft" Excel. This add-in turns Excel into a powerful OPC HDA client-reporting tool.

XML Server: Is a component, which accepts XML document requests for historical data, and via the OPC HDA server returns either an XML document or a DOM tree with the result set. Note that the XML schema is based on work in progress from the OPC XML Working Group.

Web Clients: These are web-based components requesting, receiving and displaying the XML documents via the XML Server.

A superset of RAPID targeted at the pharmaceutical industry – RAPID-Pharma – adds extra components to satisfy the needs of this industry. These are illustrated in **Figure 4.10** to the right of the vertical line, and include integrated security and electronic/digital signature capabilities (managed with the Security Manager tool), and a comprehensive Audit Trail (managed with the Audit Manager, and queried via the XML based Audit Report Tool).

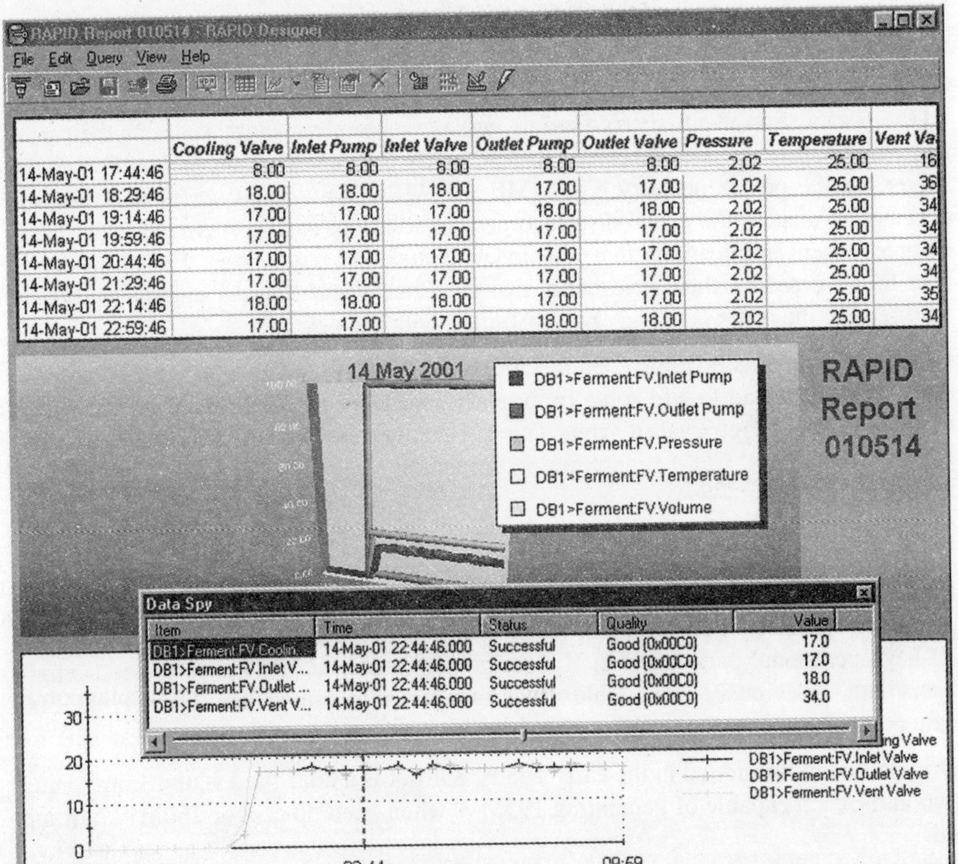

Figure 4.12: Sample Report

OODBMS technology is ideal for storing time series data as generated by process plants and offers significant performance advantages over both RDBMS and 'hybrid solutions' particularly with regard to indexing issues. RAPID uses a patent pending zero-loss data compression algorithm to reduce disk usage, and offers a fault-tolerant option to ensure zero down time.

RAPID and OPC

Standardising on OPC mechanisms for collecting and retrieving event data was a conscious design strategy when architecting RAPID.

OPC DA and A&E allowed the software development team to concentrate on writing the core RAPID components rather than having to develop proprietary driver software to connect to each SCADA, DCS or PLC. In addition customers are free to chose any OPC DA or A&E Server for their environment i.e. RAPID can connect at the PLC or DCS or SCADA level with the appropriate server. RAPID supports the OPC DA 1.0x and 2.0x specifications, and the OPC A&E 1.0x specification.

The OPC HDA Server implementation allows RAPID to provide a standard method of requesting and retrieving both raw and summary (aggregate) historical data. The HDA specification itself is very flexible and allows implementers to define their own aggregate functions. For instance, RAPID adds a number of these function to its server for the power industry e.g. TIME_AT_MIN. Raw data queries may also be executed to display the actual data recorded. In addition, customers may choose best-of-breed client applications that will interoperate with any server. The customer is also future-proofing their investment as new Clients and Servers emerge with new feature sets, the older ones may be substituted with the new.

However, as there is no equivalent of OPC HDA for OPC A&E data or for batch type data, Automsoft had to add some proprietary interfaces to our HDA implementation to allow historical retrieval of same. This is permitted by both the OPC HDA specification and standard COM programming guidelines, and does not impede on the core functionality.

Case Study (RAPID in Action)

RAPID has been generating real benefit for AES Kilroot, one of Northern Ireland's privatised electricity generators. Implemented to provide data logging and retrieval of all key operational parameters at Kilroot power station, RAPID now provides engineers with – previously unavailable – real time access to data gathered from over 1500 data points across the station's two generating units and their shared utilities.

Originally commissioned in the early 1980s, Kilroot is a dual-fired station comprising two units, each capable of generating 195MW when fired on coal or 260MW burning oil.

„Before we had RAPID, the only access to historical data was through manual paper records so this is a very significant step forward," explains Gary Shannon, the engineer at AES Kilroot responsible for implementing RAPID at the site. "Under the old system data retrieval was a centralised function. Now engineers can access historical data direct from any desktop on the plant network, which is proving to be highly effective in increasing engineers' awareness of systems flows through the plant."

AES Kilroot recognised as long ago as 1996 that it needed a computerised system of storing and retrieving process data and evaluated a number of plant data management systems before selecting and commissioning RAPID in April 1997. "None of the other products we looked at were able to mesh with the desktop in the way that

RAPID could," commented Gary Shannon. "Our long relationship and the product's flexibility have enabled specific enhancements to be incorporated into our system, as well as offering excellent development feedback to the folks at Automsoft."

RAPID runs at Kilroot on two dedicated Windows NT 4 Servers, and collects data over switched Ethernet communications links from the RTUs monitoring each generating unit. In addition about another 12 distributed I/O servers, dedicated to distributed control systems such as auxiliary steam, coal handling, water treatment and burner management, feed data into RAPID.

Logging the 1,500 data points at cycles of as little as every two seconds, the system is able to provide real time storage and direct on-line access to a complete historical record of the plant's operation over a period of six months. The current method of operating the system is to maintain a log of a full year's worth of data, six months at a time.

There are two routes to the data held in RAPID. Engineers can gain data access directly via desktop clients on the network, while the station operators in the control room have access with an ActiveX client. This access is embedded in the Wonderware InTouch HMI through which they monitor and control the operation of the plant and its services.

Because RAPID makes no assumptions about how data is subsequently to be used, its range of application is only limited by the imagination of the users. Currently, at Kilroot, applications include plant efficiency monitoring and optimisation, incident and trip investigations and emission monitoring. RAPID is implemented with a range of tools for interrogating, formatting, reporting on analysing and trending data in the database, but its flexibility allows AES Kilroot users to set their own report criteria, pulling data direct from the database into Microsoft Excel spreadsheets.

„Flexibility is the key with RAPID," says Gary Shannon. "We don't even have any predefined report templates set up. Users find it easier to set up their own reports as and when they need them. We have, on the one hand, engineers wanting to do every kind of operational analysis and, on the other, we have our business development team using the same data but looking at longer-term issues and comparing data against what are in effect design-type parameters. If you have predefined formats, you're making assumptions about what the user is actually going to need."

Bob Stuart, Kilroot's Performance Engineer and arguably RAPID's most regular user, "Typically I come in each morning and look at the previous night or weekend's data, echo Gary Shannon's views. The greatest strength of the system is its flexibility; the ability to vary the search parameters each time you access the system allows me gain the data I need at that time, and I am not fixed to an output that I then need to physically search." Bob Stuart continued, "You simply define the data points you want to look at in Historian, along with their frequency and range, and import them directly into Excel. And you can keep going back for other points as you investigate a problem. It's an excellent diagnostics tool."

Both Bob Stuart and Gary Shannon see RAPID as a huge improvement over previous methods of logging and retrieving operational data.

Ian Pepper, Automsoft International, Ireland

4.3 OPC Client Product Samples

4.3.1 SIMATIC ProTool/Pro and SIMATIC WinCC – Visualization Software with OPC Communication

With SIMATIC® HMI®, Siemens Automation & Drives is the most comprehensive provider of products and systems for control and monitoring purposes. The products offered range from simple Push Button Panels, Operator Panels with text and graphics displays, intuitive-to-operate Touch Panels, innovative Multi Panels with Windows CE, up to pure software products. The PC software SIMATIC ProTool/Pro® is employed for the on-site visualization of machines or small systems. SIMATIC WinCC®, on the other hand, is an open process visualization system (SCADA system) that enables client-server solutions with redundant servers, distributed system functions, and connection to MES and ERP solutions.

Visualization with SIMATIC ProTool/Pro

Whether for machines or in small systems, the runtime software for PC-based visualization on-site is SIMATIC ProTool/Pro®. With its exceptional performance, SIMATIC ProTool/Pro® is optimally suited for single-user systems at machines or processes. It runs on any regular PC with Windows 95 or higher. ProTool/Pro has always set the standard for simple configuration and thus short configuration times. In addition to the drivers for all major PLC types, ProTool/Pro also comes with an integrated OPC interface (under Windows NT/ 2000) that meets the Data Access 2.0 specification.

ProTool/Pro as OPC Client/Server

Via the OPC interface of ProTool/Pro, the exchange of data with different applications is made possible – particularly in networked automation solutions consisting of varied components. ProTool/Pro can be the OPC server as well as the OPC client and thus, for example, communicate with the higher-level visualization system (SIMATIC WinCC) and the automation level (PLC).

Via the integrated OPC server, ProTool/Pro supplies process data (tags) to other applications (OPC clients). Communication with other OPC servers, e.g. PLCs of other manufacturers, is made possible via the OPC client functionality of ProTool/Pro.

The exchange of data between the OPC client and the OPC server takes place directly via Ethernet with TCP/IP, or the application provides an OPC server interface – for the communication with ProTool/Pro – on the same computer, e.g. via an interface card with corresponding hardware driver and OPC server.

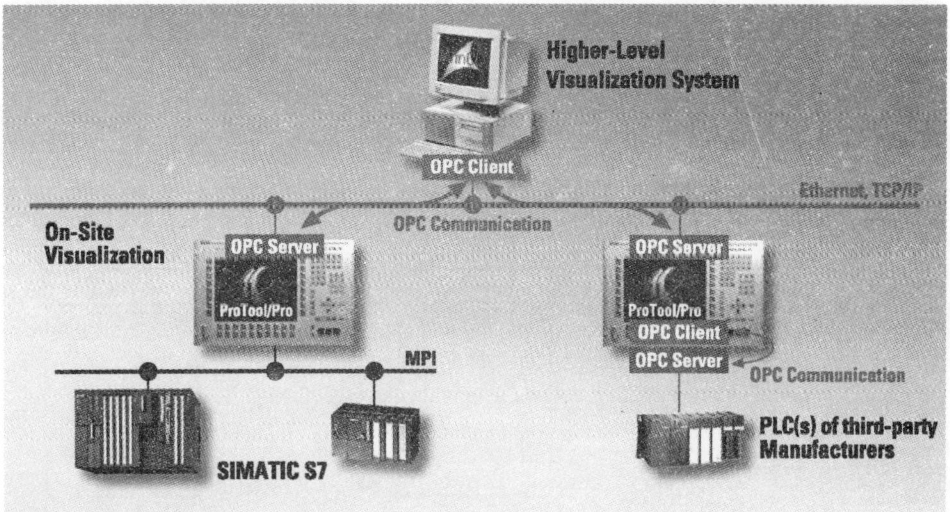

Figure 4.13: *Manufacturer-independent communication with ProTool/Pro as OPC client and OPC server*

SIMATIC WinCC – the Client-Server Solution for SCADA Applications

SIMATIC WinCC offers complete HMI functionality on regular PCs with Windows NT or 2000. Via WinCC options and add-ons as well as open interfaces (e.g. ActiveX, OLE, OPC, C-API), the user can expand WinCC to meet his or her requirements. WinCC highlights include the integrated ODBC/SQL database, the simple Wizard-guided configuration (also online), the language switch option for the configuration software and the runtime application (several languages possible), and the "Totally Integrated Automation,, approach in applications using SIMATIC. The latter stands for engineering cost savings and comprehensive diagnostics options.

The application scope of SIMATIC WinCC ranges from simple single-user solutions to complex multi-user systems with redundant servers and distributed system functions, e.g. in systems engineering.

WinCC builds upon OPC

In addition to special communication channels to SIMATIC PLCs and PLCs of other manufacturers, WinCC has been utilizing the new OPC communication standard from the very beginning. Via the integrated OPC server, WinCC makes all process data available to other applications (OPC clients). Via the OPC channel DLL (included in the basic package), WinCC can receive data from other OPC servers, e.g. from a lower-level PLC. The communication of different WinCC servers with each other is also made possible by OPC, thus allowing the implementation of cascading system structures.

This is exactly the concept that was realized at DaimlerChrysler's truck cab manufacture in Wörth, Germany. For each of five sections, a separate WinCC OPC server is employed. Positioned above the servers is a WinCC OPC client that compresses the data of the five servers and supplies it to other OPC-capable applications.

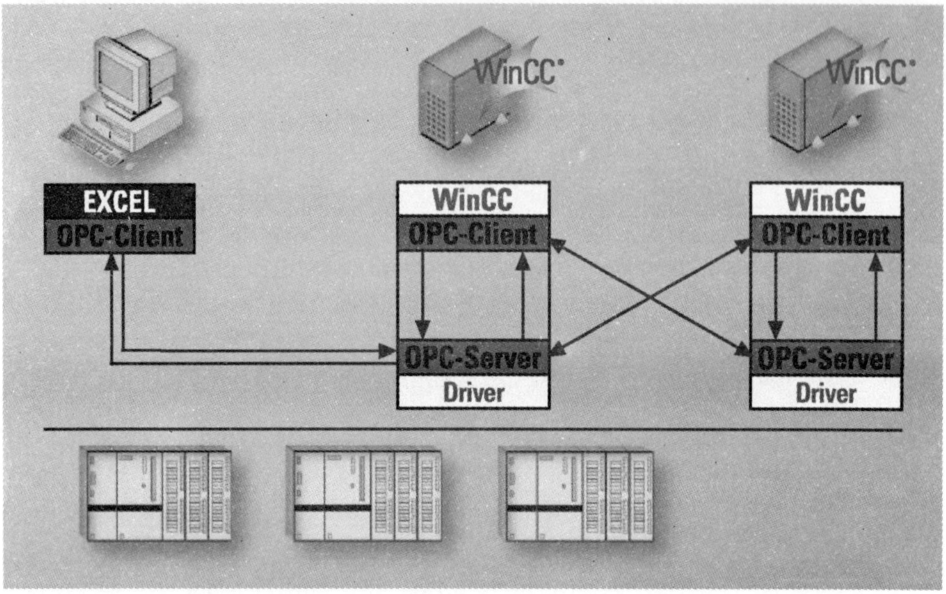

Figure 4.14: *WinCC with OPC clients, OPC servers and OPC communication to Excel*

Configuration: Easy to connect to OPC Servers

The configuration itself is very easy to perform. The configurator simply adds the OPC channel (just like a PLC channel) and then establishes the connection to the OPC server.

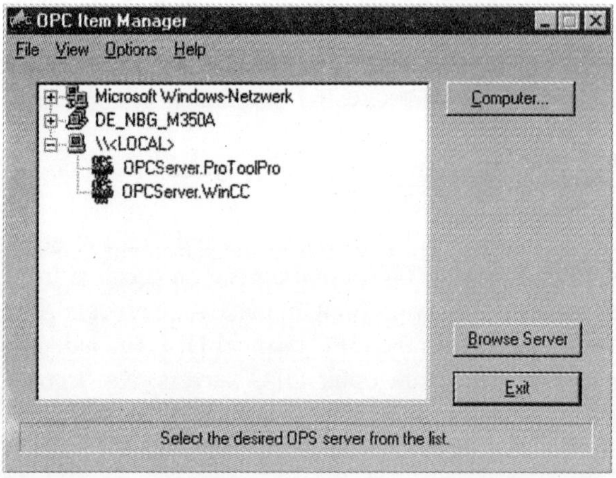

Figure 4.15: *OPC-Item-Manager for selecting the OPC server*

OPC for the Visualization: Pros & Cons

The advantages of OPC lie in the fact that it is defining an industry standard for communication between industrial field devices and software components that also applies in office applications or at the process level (MES software) and increasingly the management level (ERP systems). Whenever devices and systems of different manufacturers have to be combined in today's automation solutions, OPC is indispensable. As soon as the manufacturer of a component makes available an OPC server, this component can be connected to open systems such as WinCC.

As OPC supports a change-triggered transmission of data, an OPC connection can achieve almost the same performance as a proprietary driver solution. As OPC not only supports access to individual data points but also the transmission of alarms and archive data, it fulfils the essential requirements of a visualization solution.

Not yet solved to satisfaction is the behavior following an interruption of the DCOM connection between client and server. Already, detecting such an interruption can be challenging, as the protocol stack requires too much time (due to timeouts and retransmissions common with standard network protocols) to report an error to the application program. Thus, the user is informed relatively late that a displayed value is no longer valid. In such instances, powerful analysis tools would be beneficial.

As the access to complex data structures (arrays, structures) has not yet been standardized, different implementations at OPC-capable products can still be found. However, this will be resolved with a soon to be introduced standard.

Manfred Baron, Siemens AG, A&D PT1, Nuremberg, Germany

4.3.2 DIA*dem*® as OPC Client

A pressing need in both factory automation and process manufacturing is making production and business systems work together. Often this has been a barrier because of incompatible proprietary communication interfaces from different suppliers of automation hardware and software. OPC gives production and business applications across the manufacturing enterprise access to real-time plant-floor information in a consistent manner, making multi-vendor interoperability and "plug and play" connectivity a reality.

National Instruments, a leading supplier of computer-based measurement and automation products, is committed to the development of state-of-the-art technology and open industry standards like OPC that improve productivity for many users. Therefore, NI is able to provide users with many OPC compatible products such as LabVIEW, Measurement Studio, DIA*dem*®, FieldPoint, and others.

This chapter gives a short introduction to DIA*dem*® and provides an overview of how DIA*dem*® employs the OPC standard for different applications. In principle this could also be demonstrated using other National Instruments software such as LabVIEW or Measurement Studio.

The main reason why DIA*dem*® is employed in so many areas is its consistent use of standardized interfaces: In the process industry, fieldbus systems are monitored and online events are logged via OPC interfaces, whereas in the research industry OPC interfaces serve for the automation and analysis of pilot plants and test stands. DIA*dem*® serves as a key element: OPC enables the connectivity between fieldbuses such as AS-interface, CANopen, DeviceNet, Modbus, Foundation Fieldbus, Interbus, PROFIBUS, P-Net, and others. At the same time – due to the driver library integrated into DIA*dem*® – different plug-in boards and measurement devices can be used for direct data acquisition. For process visualization DIA*dem*® offers online instruments as well as an extensive symbol library.

Easy Access to Automation Data

Connecting DIA*dem*® to existing OPC servers is straightforward: in the software device (DAC), which is used to define measurement tasks, the user selects the OPC block and opens a browser by double-clicking it. The user is then offered a selection of OPC servers on his own PC (local host) or in the network (**Figure 4.16**). He can choose between two methods of searching for servers using the OPC standards 1.0A and 2.0. Once the server is selected, the user is shown a dialog box similar to the Windows Explorer in which he can select the items (measurement nodes) offered by the server. Now the user can start the measurement; measured data are presented as curves, numbers, or bars.

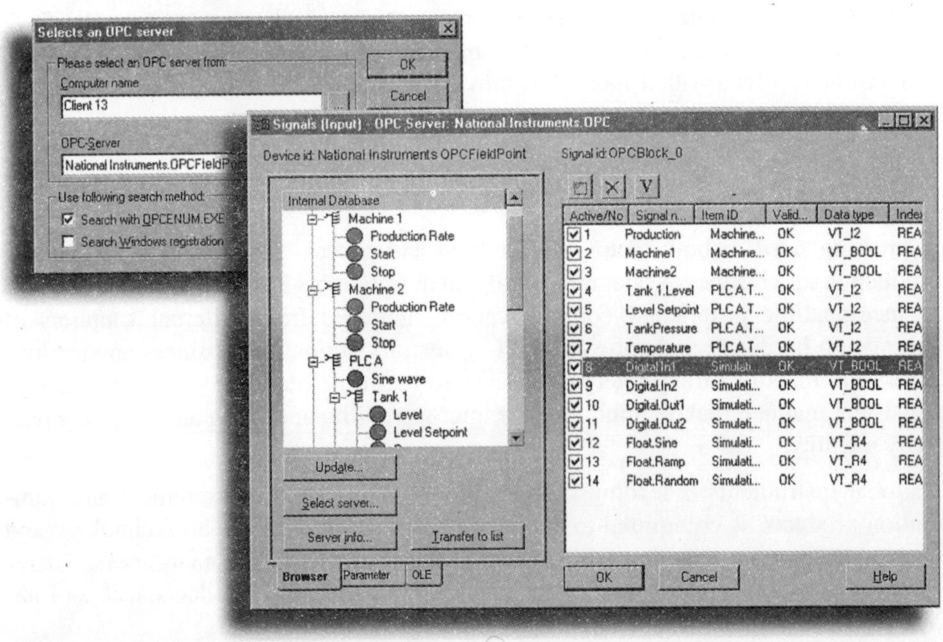

Figure 4.16: The OPC client of DIAdem® scans OPC servers on the own PC or also in the network via two different search methods

In practice, further configuration possibilities at the OPC client level proved to be useful. The user can select the synchronous or asynchronous measuring mode and the timeout preferences can be specified for the OLE communication. Log files and error handling facilitate the removal of possible communication problems between client and server.

Monitoring and Reporting Processes

Currently, OPC is mainly used for the connection of fieldbuses and automation components such as in the process and manufacturing industry. Due to DIA*dem*® being OPC compliant, its data analysis and documentation capabilities can now also be used in these areas. The automatic generation of logs and reports in DIA*dem*® is particularly appreciated by users, who – due to prevailing quality standards in the industry – must present meaningful documentation of their measurements. In this context, it is worth mentioning the alarm and logging system in Dia*dem*® that can be used for monitoring and documentation tasks and that saves the subsequent manual analysis of measurement data. Fieldbus devices, PLCs, and measurement nodes acquired via OPC are monitored on critical events; alarm messages are generated and logged in a log file. The historical trend of the measured data can be viewed on the screen. The user can enter comments and additions online (**Figure 4.17**). An alarm table lists the current messages and allows authorized users to acknowledge alarms. This, too, is documented in the log file, so that user actions can be reconstructed at any time.

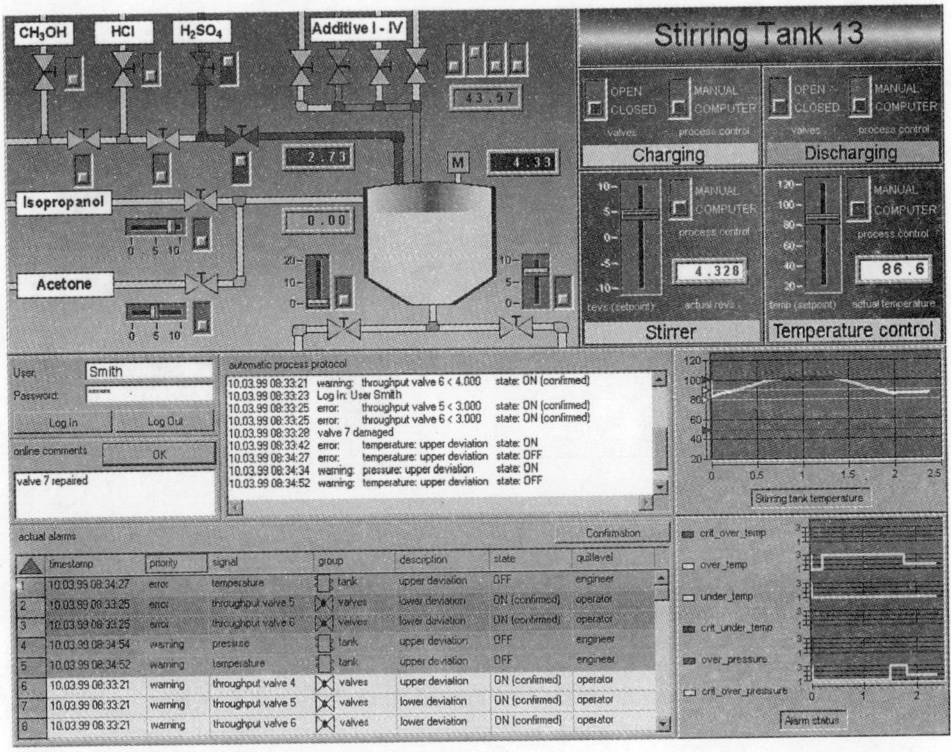

***Figure 4.17:** DIAdem® integrates automation, visualization, archiving, and user administration as well as an alarm and protocol system.*

The alarm and logging system of DIA*dem*® also serves as communication tool: important messages can automatically be sent via fax or e-mail. If the DIA*dem*® PC is integrated into a computer network (Ethernet), messages can be exchanged with a central monitoring PC via TCP/IP. This way, decentralized systems are monitored and documented from a central point.

Evaluating and Documenting at the Touch of a Button

All data acquired with DIA*dem*® can be directly analyzed and presented as process or test reports. This can also be handled by a second (networked) DIA*dem*® computer, even as the data is acquired. Thus, analysis algorithms programmed by the user are no longer necessary. All statistic functions needed for the analysis of technical data are contained in DIA*dem*® and can be used immediately without having to be programmed by the user. The raw and processed data are presented as tables or diagrams (**Figure 4.18**).

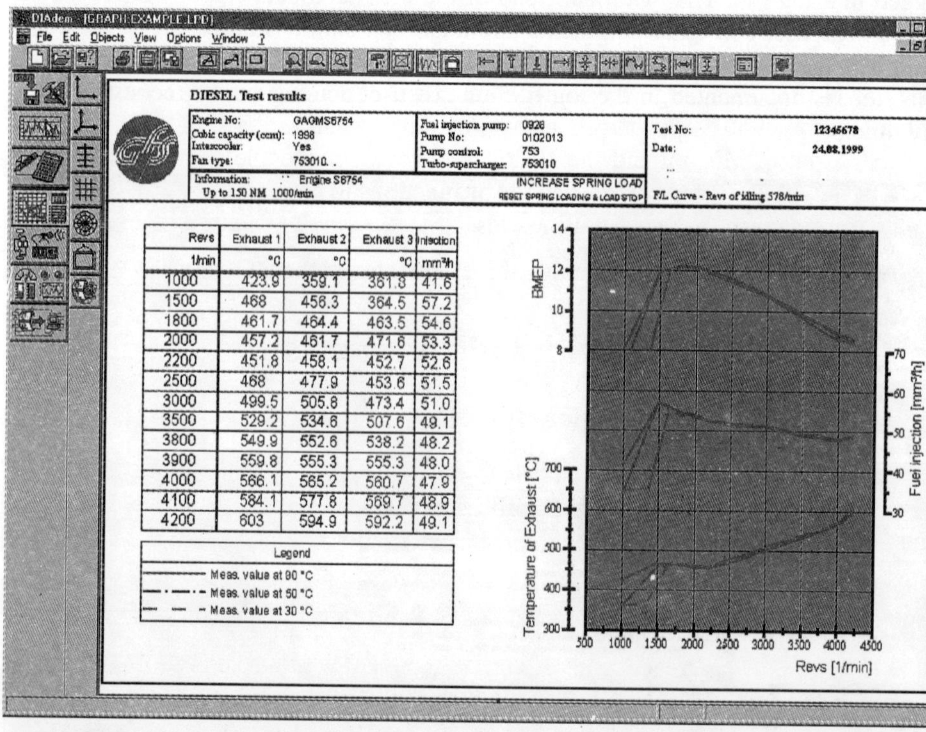

Figure 4.18: *Process and laboratory data are analyzed and documented with DIAdem®. With the help of auto sequences these analysis processes can be handled automatically – without being programmed by the user.*

DIA*dem*® is easy to use: With the help of graphical dialogue editors, parameter input masks are generated that can parameterize measurements, control sequences, and analyzed data – without any programming being necessary. Thanks to the teach-in mode, auto sequences – the macros of DIA*dem*® – are automatically generated, once the user has shown the program how to do this. This way, even complex applications can be developed comfortably and within a short period of time.

Seamless Integration

Not only do users benefit from the implementation of OPC into DIA*dem*®, because the PCs are much more easily connected to PLCs and fieldbuses. Also National Instruments as a software manufacturer profits from the OPC standard: the development of drivers for an enormous amount of fieldbuses and controls becomes superfluous. Hardware manufacturers recognized this a long time ago and offer OPC servers as "drivers" for their hardware. The benefit of the standardization and interchangeability of measurement data made possible by OPC is obvious: now there is one uniform interface for all software and hardware components within the area of industrial automation. This compatibility of software and hardware products, called interoperability, enhances the user's productivity in regard to application development enormously and results in immediate cost reduction. Thanks to uniform interfaces, maintenance costs can drastically be cut down too.

For applications requiring higher channel count, distribution, vision, motion support, and other features, LabVIEW and its add-on modules (such as the Datalogging and Supervisory Control Module and the Vision Development Module) provide a complete solution. DIA*dem*® provides a rich set of functions to connect to LabVIEW and allows the user to take the best of both worlds.

Dieter Peuser and *Rahman Jamal*, National Instruments, Aachen, Germany

4.4 Real Applications with OPC Products

4.4.1 OPC at Tetra Pak Converting Technologies AB

Tetra Pak Converting Technologies AB is a core competence company within the Tetra Pak group that develops converting processes for Tetra Pak's different packaging systems.

The company has about 100 employees and is located in Lund, Sweden. The customers are the Tetra Pak packaging systems companies and converting factories.

Systems Integration is a department within the company whose role is to find, integrate, and implement new automation and industrial IT technology.

The first ideas

The first time our company came across OPC was late 1996. We had for a long time said that we should try to standardize our communication structures as far as it was reasonable, both from a development standpoint and a maintenance standpoint. OPC seemed to be a standard that at least could help us to hide the underlying physical protocols from everyday exposure. The more we looked at it, the more attracted we were, as we realized that we could share information from the factory floor much more easily and freely with any interested partner.

The first project

Persuading the customer

In January 1997 we got the first opportunity to start a customer project with the new technology, porting an existing production data feedback system from VMS to Windows NT. It took some time to persuade our customer to be a pilot and to invest in a technology that wasn't established yet. But when we showed them what significant savings they would make on the total expenditure, and the increased flexibility they would get, they agreed that it would be a much better solution.

Development

There were a lot of unknowns in the project. We decided that in terms of OPC, we would buy the server from an external supplier, but develop the client internally as an integral part of the data acquisition system. We looked around for available OPC servers, but really did not find anything. We therefore made contact with a number of potential suppliers to see if we could interest anyone for a joint venture. Softing turned out to be the partner we were looking for. They developed the OPC server based on a joint specification while we helped to test it thoroughly in our lab with our internally developed client.

Installation

In May 1997 we were ready to do a test installation at the customer site. We were ahead of schedule and below budget. In June we did the installation. Basically every-

Figure 4.19: *Converting machines laminating kilometers of paper coils*

thing we installed was either a beta or a pre-release product. In a week we connected five machines and had all the production feedback functioning properly.

The system was left running for three months as a run-time debug test. It crashed a couple of times, giving us enough information to make some necessary modifications to the software.

In September 1997 we returned and finalized the installation. We have heard no problems from the customer since.

The solution

In short, the new structure consists of one small dedicated Siemens S7-215DP PLC for digital and analogue signal collection plus one Serial-to-PROFIBUS converter for bar-code data per production machine. An optical PROFIBUS DP factory network is used to convey collected data to a centrally located NT server. The NT server has a PROFIboard and an OPC server from Softing and acts as PROFIBUS master, polling the machine units for data. The internally developed software contains an OPC client that gets all the PLC data from the OPC server, builds, and sends messages to an administrative production system.

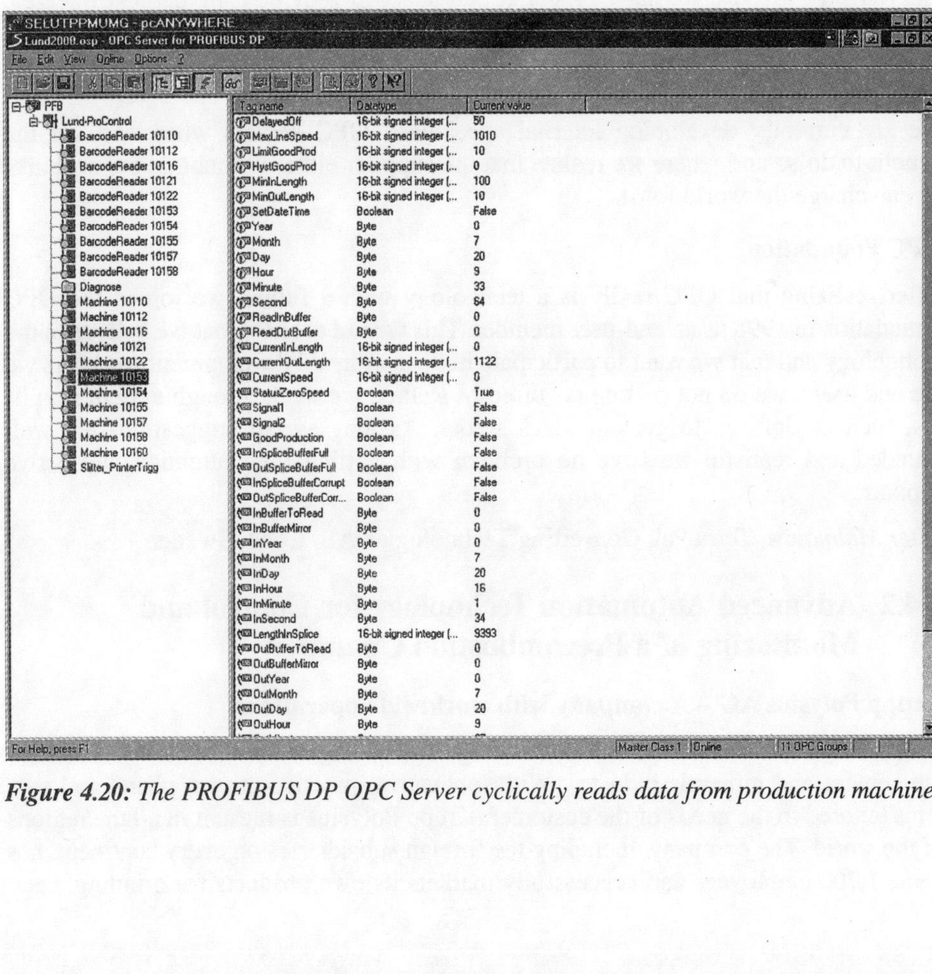

Figure 4.20: *The PROFIBUS DP OPC Server cyclically reads data from production machines*

Status today

As of July 2000, this data collection system has been installed in ten factories. Over the next couple of years we estimate several more installations.

Current projects

OPC/XML

With the success of the first project and the overall awareness of OPC as an enabling technology, we are encompassing OPC in a much wider scope. We are currently developing a completely new framework for future applications where OPC and XML are cornerstones. Naturally user interfaces will be Web-based and available for a whole range of new devices.

External OPC Servers

As we see OPC as a preferred way of communication on the factory floor, we encourage our equipment suppliers to provide OPC server functionality – not only for Tetra Pak, but as a technology that makes commercial sense for the general customer. The most common supplier reaction is slowly changing from a confused "OP.. what did you say" to "yes, we have looked at that and will provide it in the next release". Our job as evangelists is becoming easier.

Internal OPC Servers

We are currently developing internal specialized OPC servers where we see the benefit to do so and where we realize that the supplier either will not do so, or thinks he can charge the world for it.

OPC Foundation

After realizing that OPC really is a technology with a future, we joined the OPC Foundation in 1998 as an end-user member. This we did to show that we believe in the technology and that we want to participate in the continued development work. As we are end users, we do not participate directly. Rather, we work through system suppliers, such as Softing, to get our ideas across. As long as our suggestions are well founded and realistic we have no problem with getting their attention and active support.

Peter Malmqvist, Tetra Pak Converting Technologies AB, Lund, Sweden

4.4.2 Advanced Automation Technology for Control and Monitoring of a Precombustion Chamber

Krupp Polysius AG – A company with worldwide operations

Krupp Polysius is one of the world's leading engineering companies for equipment in the cement and minerals industry. With innovative, practice-proven plants and systems tailored to the needs of the customer, Krupp Polysius is present in all the regions of the world. The company, including the foreign subsidiaries on every continent, has some 1,700 employees and successfully markets its own products for grinding, heat, materials handling, and automation technology. Through its own professional project and handling management, Krupp Polysius handles the project planning, design, delivery, erection, and commissioning from the individual machine through to the complete plant. The services of the company are not purely limited to the planning and construction of new plants; they also include the modification and modernization of existing plants with the aim of reducing operating costs and meeting the growing demands of environmental protection.

Introduction

As part of the order to install a precombustion chamber in a cement works in Switzerland, a program for process control and visualization was developed. The plant is controlled by a process control system. The Polysius system for process control is based on fuzzy logic and performs super ordinate set point control. The OPC technology was chosen and integrated into the process control system as data interface to the PCS and the process periphery. With this standardized open system technology, communication with any other peripheral systems (control systems, process control technology) will also be simplified for future orders.

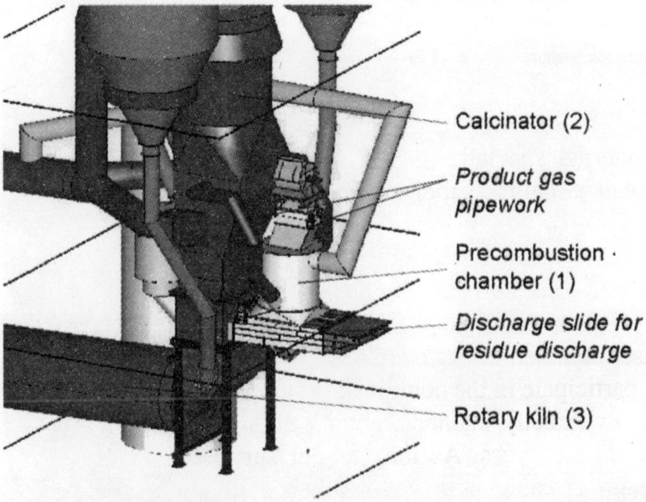

Figure 4.21: *Schematic diagram of the precombustion chamber with kiln system*

The precombustion chamber (1) is a process engineering installation for generating energy from uncrushed, coarse secondary fuels. The objective of the process is to cost-effectively replace a large proportion of the conventional fuel used to date for precalcining, e.g. coal, with secondary fuels – in this concrete case with old tyres and waste wood. The process in the precombustion chamber supplies a product gas and fine-grain solid products. These products exhibit flying and burning properties similar to those of conventional fuels and can thus be fired directly in the calcinator (2) of the kiln system. The non-combustible reaction products such as the tyre cord wires or mineral fractions in the wood are admitted to the rotary kiln (3) gradually via discharge slides and bound into the cement clinker by the subsequent sintering process.

The process thus operates without residues. The plant is characterized by low processing and operating costs with a high specific throughput.

Measurement of the fuel density in the precombustion chamber

The reaction zone of the precombustion chamber contains hot fuels and hazy product gases. The task of not only detecting limit states but of also obtaining a differentiated picture of the chamber content is solved by means of a radiometric filling level measurement. The measuring system consists of a radiation protection vessel with cobalt preparation and two scintillation detectors with the evaluation units. The radiation source is mounted on a height-adjustable device and is moved up and down along the reactor wall in the relevant area for measurement. The detectors are mounted accordingly in the same height range on the opposite wall of the reactor. Depending on the position of the radiation source, the outlet angle means that only a few of the detectors are irradiated. The measured radiation strength or pulse rate is a measure of the local attenuation of the measurement signal and thus gives an indication of the mean density of the fuel density at the specific height in the reactor. From the measured pulse rates, a density profile is calculated cyclically over the height in the reactor from which conclusions can be drawn as to the structure of the fuel charge.

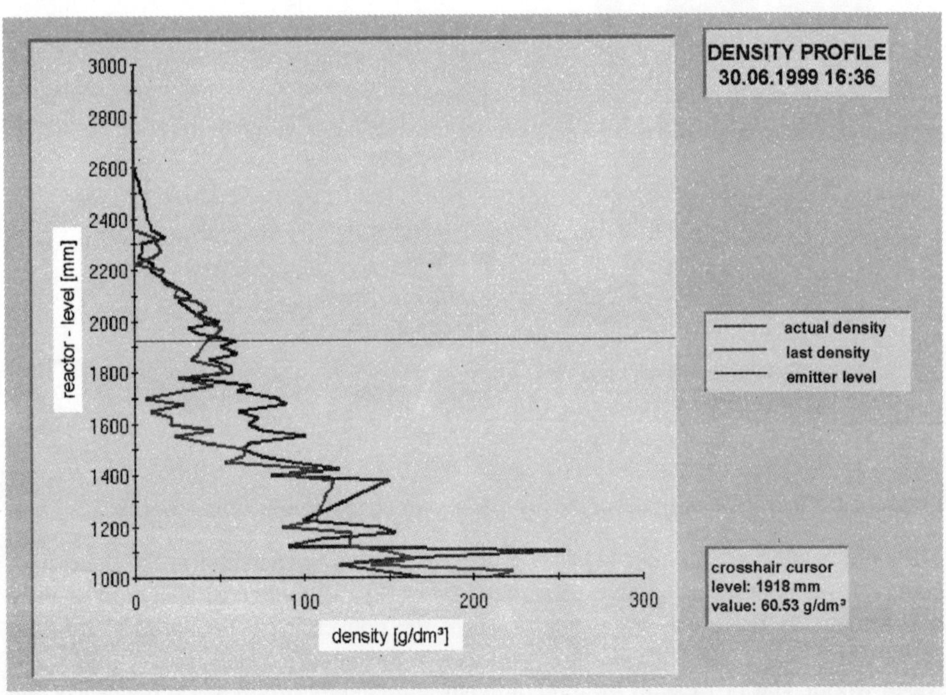

Figure 4.22: Visualization of the density curves

After every measurement cycle, various parameters are determined automatically from the profiles and visualized, i.e. the fuel mass in the reactor (mass), gradient parameters, and deviations from a reference profile (evaluation). The calculated parameters are input to an "interpreter" which generates an output value from these input values. This characteristic value represents the filling level of the reactor. The filling level interpreter was created using a fuzzy logic module. The following 3D diagram shows in simplified form the transmission characteristic of two input parameters and one output parameter of the fuzzy system.

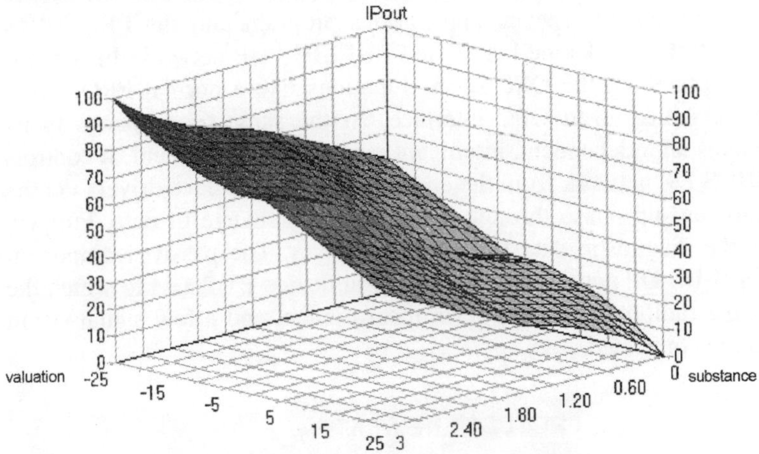

Figure 4.23: *3D characteristic of the filling level interpreter*

Automatic control of the plant

At the end of the start-up procedure, i.e. when a more or less steady state of the plant is reached, the plant is operated automatically by the process control and regulating system. The fuzzy logic-based superimposed control system from Polysius performs the set point control. Here the incoming information, i.e. filling level, charge volume, and product gas temperature, are processed and converted into manipulated variable data which are transmitted to the control level.

For the precombustion chamber, different control strategies are employed for the operating modes tyres and wood. In "tyre" mode, the height of the residue bed is held constant via the speed of the discharge slide and the chamber temperature is controlled via the admitted hot air volume. In extreme operating situations, the charge volume is also modified. In "wood" mode, the discharge slide is not available as a control element. In this case, the bed height is held constant via the admitted hot air volume. A further parameter is the specific flow resistance of the wood filling, which can also be influenced, if necessary, via an adjustment of the fuel volume. The information on filling level (characteristic value, IPout), chamber temperature (e_TPG), and bed resistance (wood only) are related to the output parameters hot air volumetric flow (lambda), discharge slide speed (vSch), and the set point change of the fuel mass flow (deltaMBr) in a fuzzy logic module.

Open system structure for automation products from Krupp Polysius AG

The process control system from Krupp Polysius consists of a PC with Windows NT operating system, a PCI board as PROFIBUS master interface (PROFIboard), and PROFIBUS DP slave modules for the digital and analogue input and outputs.

The program for automatic evaluation of the filling level profile includes the visualization and archiving of the process data, fuzzy logic modules, and the OPC client as data interface. The system also functions as a superimposed controller, which generates manipulated variables from the filling level profiles using the calculated characteristics; these variables are transmitted to the control system. An OPC server acts as an intelligent link between this special applications program and the PROFIBUS master. The master interface links the PC to the PROFIBUS DP network. In contrast to the OPC client application, the OPC server functions like a normal PROFIBUS device driver with a standardized OPC interface. At the same time, thanks to its functions, it represents an independent, intelligent automation component. It controls the whole PROFIBUS DP network from the configuration of the slave drivers via the communication and start-up phase through to the cyclic exchange of data. Furthermore, it evaluates the diagnostic data from the DP slaves. The passive component (slave) in the PROFIBUS DP network is the peripheral device ET 200M to which the process signals of the radiometric filling level measurement and a few signals from the control system are connected.

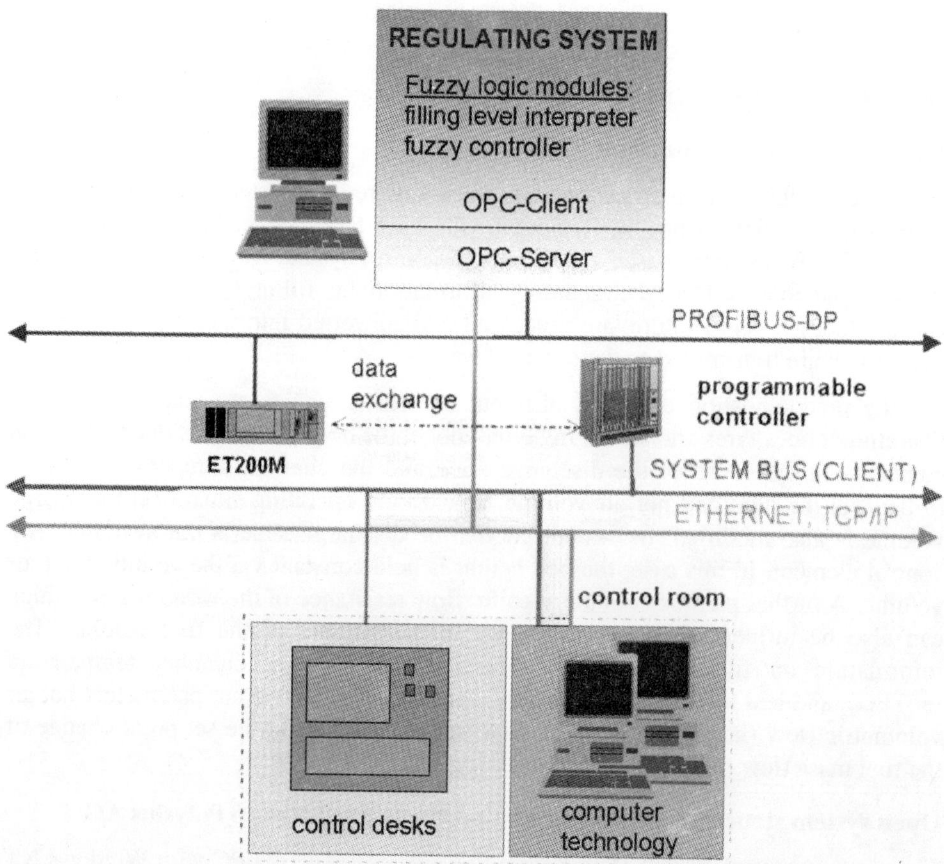

Figure 4.24: System structure

The data between the control technology and the process control system is currently exchanged via the analogue and digital input/output modules of the bus systems (PROFIBUS DP, system bus). The control technology (control desks, relay technology) is gradually being converted to modern computer technology with Windows-based applications software. The process control system communicates via a TCP/IP network with the control system. The integral OPC client of the control system does not have to be modified for this, as with the DCOM expansion of OPC, the client can already access the data on a server (here a control system) via a network. The remote computer of the control system merely has to function as the OPC server.

The OPC interface was integrated with the software of the process control system using the OPC client toolkit from Softing. The software was developed with MS Visual C++ to be consistently object-oriented and is freely configurable with respect to the communication partner and the data to be transmitted. The software now serves as the basis for all Polysius application programs, which handle the data exchange with the Polysius control technology POLCID® NT and laboratory automation POLAB® AMT or with external systems.

Previously, different communication concepts were pursued in data exchange between Rockwell software-based Polysius automation products POLCID® and POLAB®, analyzers, low-ranking controllers, industrial PCs and external devices (i.e. serial interfaces), DDE, file transfer and system-specific interfaces (e.g. ROI/Rockwell).

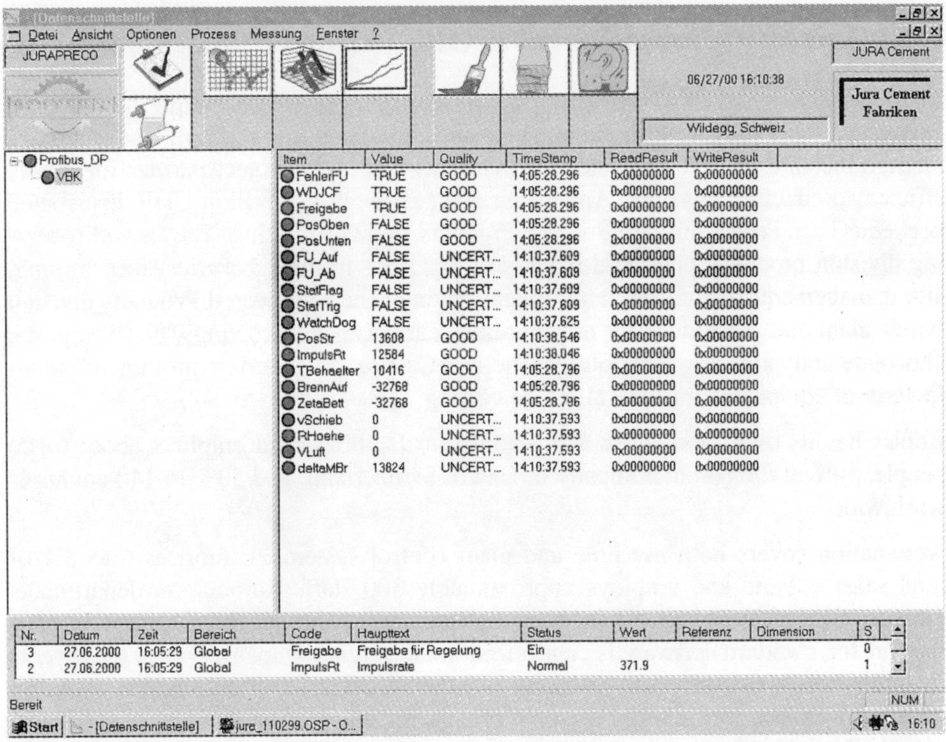

Figure 4.25: OPC communication mask

The ROI interface from Rockwell was no longer able to meet the growing demands of POLCID® and POLAB®. The DDE data transfer has already been superseded by Microsoft OLE and is no longer supported. Every hardware component generally still had its own driver, which did not implement a standardized interface. From the point of view of the Polysius application programs, no standardized access to the real process data was possible. The communication technology OPC now offers the possibility of standardizing the access to the process data. The unhindered and transparent data exchange between the automation programs and all other Windows applications now clears the way for a break away from the hybrid control and EDP systems and a move towards homogeneous factory automation.

The Polysius application programs, which act as OPC client, contain a comfortable display of the active connections to the OPC server. The name of the OPC server to which a link has been set up and the transmitted values are displayed in a separate window. The time of the last update, status, and value of the process parameter are automatically updated when a change occurs. Faults in the connection setup or communication are signaled optically.

The around 20 Polysius program applications with the implemented OPC interface in use in the meantime are very easy to maintain, as all the versions are based on the same concept.

Dipl. Ing. *R. Teutenberg,* Krupp Polysius AG, Beckum, Germany

4.4.3 Increasing Employment of OPC in Buhler Plant Control Systems

Buhler AG has a long tradition in machine and equipment manufacture, and is today a modern technology and knowledge provider as well as a project partner for highly efficient production systems. Annual turnover is about 1.4 billion CHF, distributed between Grain Processing, Engineered Products, and Die Casting. The Grain Processing division produces plant and machinery for flour milling, bakeries, feed milling, silo management, rice milling and optical sorting. The Engineered Products division builds plant and machinery for printing inks, chemicals and coatings, plastics, pasta, chocolate and extruding technology. The Die Casting division is a prominent manufacturer of equipment for light alloy die casting.

Buhler has its own fabrication and automation facilities, and employs about 6,100 people, 50% at the parent company in Uzwil, Switzerland, and 50% in 140 countries worldwide.

Automation covers both machine and plant control systems, comprises 6 to 8% of total sales volume and employs approximately 300 staff. Automation departments exist in all the divisions and in many subsidiaries, although development of, and support for, standard software is centralized in the parent company. Due to the diverse requirements of the different divisions and the varied level of automation found in different locations worldwide, the requirement for automation systems is extremely varied. The following article describes the experience gained with the introduction of

***Figure 4.26:** Flour milling machines*

OPC to Buhler AG and discusses the strengths and the weaknesses of the technology, as well as the future potential of OPC for the Buhler organization.

***Figure 4.27:** The control room of a pasta production plant*

Previous Use of OPC

The first contact with OPC technology came about from the use of various process visualization packages in conjunction with PLC software from Rockwell. This work was undertaken by automation departments in our subsidiaries operating in the English language areas, not by the central research and development department. The feedback was mostly positive, although initially problems were reported with access speeds.

The first use of OPC in a standard software product was in the plant control and automation system PIC (Process Interface Controller), which was developed in 2002. PIC is a software package of medium complexity, composed of third party process visualization software and internally developed program code, based on standard Microsoft components. The features of both a typical control system and those specific to PIC are summarized below.

Characteristics of a Typical Buhler Plant Control System

- The materials handled are mainly bulk and dense-flowing (dough, ink, chocolate, oil).
- The processes include requirements to transport, store, dose, mix and grind these materials.
- The plant may include both continuous and batch oriented processes.
- Most processes are relatively slow.
- Most signals are digital.
- There are in the range of 100–1000 input/outputs (maximum 5000–10'000).
- A conventional control architecture is used: PLC sequencing and PC supervision
- A high degree of functionality is incorporated at PLC level (more reliable and robust).

The above characteristics typically apply to systems implemented by Grain Processing and Engineered Products, where most plant control systems are sold.

Characteristics of the PIC Control System Software Package

- Provides integrated process visualization (HMI/MMI) and MES functionality.
- Uses standard third party products for process visualization (WinCC, RSView, InTouch).
- Includes MES functions (automation part), implemented using Visual Basic and MSDE/SQL Server.
- Supplies core functions including: silo, material and recipe handling; job-oriented production control with job lists; the storage of production data including product traceability.

PIC is distributed to the various Buhler divisions as a partially open software system. A common core is maintained centrally, which is then adapted by the various automation departments to their own process-specific requirements.

Requirements for OPC

The Buhler standard software supports both the Siemens and Rockwell (Allen Bradley) PLC families. OPC functionality was added to PIC for communication between the PLC and PC software layers. With the introduction of the new generation of Control Logix PLCs, Rockwell switched from manufacturer-specific interfaces to OPC. For us, this generated the requirement for an OPC interface with the following features:

- OPC for use with in-house automation software (MES functionality, not visualization)
- Robust and established technology for deployment world-wide
- Functional compatibly with the existing interface between PLC and PC
- Medium-sized data sets, between 100 and 1000 data points
- Coherent read and write operations at the beginning and end of jobs and batches
- A programming interface (API) for Visual Basic
- A solution that is understandable by programmers in the various divisions.

Specific Solution with OPC

We built an interface for the Rockwell OPC server into our standard software. We retained the original block-oriented communication for Simatic S7, as Siemens did not, at the time, recommend the use of OPC, for performance reasons.

We integrated the Rockwell RSLinx OPC Server (OPC V 2.0) using Ethernet as the transport medium between the PLC control software and PC supervisory level. The ability to read and write coherent data sets is not provided for in the DA protocol, so we used our own additional handshake mechanisms and the synchronous operations of OPC (Read from Device).

The software was implemented using Visual Basic calls, although this caused the restriction to use the automation interface.

Compatibility with the existing software interface was achieved, using the convention that the names of the OPC Items must be the same as the symbol names used in the PLC programs. Because the existing interface is standardized and for the most part follows a consistent naming convention, the OPC Items can be generated automatically at PC system start-up, using these rules and the existing plant configuration data.

So far more than 10 projects have been implemented using OPC. The number is relatively small, since the majority of our plant control systems are implemented using Siemens PLC technology. The largest project using OPC is a bakery control system with about 700 input/output signals. The OPC interface between PLC and PC includes approximately 8000 OPC Items in about 120 groups.

Positive Experiences

The biggest advantage came as a result of the transition from absolute addresses to symbolic names. This change could be introduced simultaneously on the Control Logix PLC and in the OPC interface on the PC. The positive points from it are:

- The decoupling from of absolute addresses (N-File and Offset)
- A more understandable interface, owing to the use of symbolic names
- A more expandable interface, without constraints of absolute addresses
- The Tag browsing facility

Negative Experiences

The negative points can be regarded to a large extent as teething troubles. Each new technology has its strengths and weakness, and one must learn to recognize them in order to deal with them efficiently. This learning process was no more difficult with OPC than other new techniques. The most serious negative consequences were as follows.

- Performance problems were encountered whilst reading and writing coherent data blocks.
- The OPC server software layer is felt to be large and rather obscure, particularly with regard to maintaining runtime performance and data consistency.
- OPC know-how is still not commonplace, even with large manufacturers such as Rockwell.

From the negative points we can highlight the features we would most like to see included in the future development of OPC:

- Improvements in both the performance and the implementation of the OPC servers.
- Better facilities for handling structured data, with provision for user-defined structures and methods to transfer data in such structures to and from the process level as complete entities.

Evaluation of Future Development and Use of OPC

Buhler is primarily a system integrator, and therefore a user and not manufacturer of OPC components. We can certainly profit from a uniform standard, because we use automation components from different manufacturers and we would be happy to adhere to such a standard in the development our own program components.

Unlike OPC users from the ERP world, we are concerned with near real-time automation. For us, the question of performance is very important. Extensions for the IT and ERP worlds, like OPC XML, are of secondary importance to us, although nevertheless a welcome addition.

Our prognosis for the employment of OPC is to a large extent positive. OPC will become the preferred solution in our new plant automation systems. With reference to our two main suppliers of automation components, Siemens and Rockwell, there is now improved support for OPC from Siemens. This will enable us to overcome the current difference in the approach we now use and, in the future, communicate with both PLC families using OPC.

Our need for OPC components will increase. As a system integrator we will, if possible, buy OPC components, and we will generally use the OPC servers provided by the PLC manufacturer. The mixing of OPC servers in a single application seems to function and we trust the manufacturers to offer well-optimized OPC servers for their own devices.

Where we develop program code ourselves we use neutral OPC program packets. Our minimum requirement is for OPC client programs. We are currently discussing the possibility of providing data from our process control PC's via OPC. In this case, we would use OPC server toolkits as well.

We see the adoption of further OPC protocols as substantially more problematic. Today only OPC Data Access (DA) is in common use. Special protocols such as alarm/event, batch and historical data are defined, but hardly used. These protocols are therefore not attractive for us, because we cannot assume that they will not quickly disappear from the market. OPC XML is, in our opinion, a different case. Here it concerns another representation of the data and offers independence from Microsoft OLE/COM/DCOM. Both points are relevant for us, but not urgent.

In general we see the variety of OPC standards as a problem, and this point was also recognized by leading opinion in the OPC world. Our ideas for the advancement of the OPC standards go in the following direction:

- The separation into 3 different aspects, which can develop independently: data structures, the use of data and the representation of data.
- Generally fewer but more comprehensive standards.
- The aspect of data structures should be based on the well-known hierarchical tree format (Items, Groups etc), supplemented by additional mechanisms. Such additions are structured data types (Complex Data) and access functions for multiple records (aspects from Alarm/Events, Historical Data). From an operational point of view the functionality of Data Access (DA) and Data Exchange (DX) should merge.
- The aspect of the use and meaning of data should be separated. Here I imagine a separation as with XML, where the basic standard is neutral. The use of Tag Names is on a second layer, fixed through the XML Vocabularies. Thus the actual data can be modeled by areas such as Alarm/Event, Batch and Historical Data etc.
- The last aspect is data presentation. This can be selected independently of contents, like in Unicode (UTF-8, UTF-16) or SOAP (XML, binary). The user interface (API) can also be bound to this format. Currently this means that there is a representation of OPC for Microsoft COM and a representation in XML. A further possible representation would be an optimized format for Microsoft NET. At the moment one can depend on .NET and COM compatibility, but in the medium term COM will become less important.

Conclusion

Despite various difficulties, OPC remains the only realistic standard for access to process data, and our planning is therefore based upon OPC. As system integrators we are keen that a uniform standard spreads and is developed further and we wish OPC every success in achieving this goal.

Fritz Langenegger and *William Conrad*, Buhler Group, Switzerland

4.4.4 MCD and OPC – two plus points for automotive manufacturing

Electronics onboard – a dilemma?

A simple case of jumping into the car, turning the ignition key to start up the engine to get from A to B? That is just not enough for a lot of drivers nowadays. Doors are unlocked with a simple touch of the hand, the trunk conveniently opens at the push of a button for your golf clubs and sometimes the doors shut as if by magic. While your seat silently glides into your own personally defined position, you can quickly take a look at what is going on on the stock exchange via the integrated onboard monitor. Lightly touch the starter switch and your engine is raring to go, and without the driver having to do very much, the vehicle is cleverly navigated around two traffic jams. A not insignificant number of small computers (referred to as *ECUs* in the automotive sector) takes care of sticking to the laws of physics while the lady behind the wheel – already somewhere on the green in her imagination – has just decided to try out a 3 iron at the first tee-off.

The countless comfort and safety features that make for relaxed driving are more often than not realized as software. And the state-of-the-art multimedia features of today's top-of -the-range vehicles would be inconceivable without extensive software engineering. These software achievements, however, not only provide a number of obvious advantages; they also result in a number of undesired side effects. It is common knowledge that software is prone to errors. As far as customer satisfaction is concerned, software errors are slowly driving the "classics" from their top positions: an engine that only starts under protest in the winter is probably slightly less annoying than having to wait days for the repair shop to call to say there is nothing actually wrong with the vehicle. The crazy blinking of the various warning lamps – just a software error!

What does that mean for a supplier to the automotive industry? It certainly means that ECUs have to be tested efficiently to ensure they fulfill the hard, economic demands of the market as well as to provide maximum security against errors.

In this context, testing, or diagnostics, means accessing the ECU functions as well as connected sensors and actuators. This usually takes place via the bus interface of the ECU, e.g. via a CAN or LIN bus. To be able to use the sensors and actuators of an ECU, the test system has to be able to access both the bus signals transmitted cyclically (' onboard communication) and the functions of ECU diagnostics (' off-board communication). This accessing of diagnostic functions usually takes place using a subordinate communication protocol such as *KWP 2000* or *UDS* based on a transport protocol such as *Diagnostics On CAN* or *K-line* (see **Figure 4.28**).

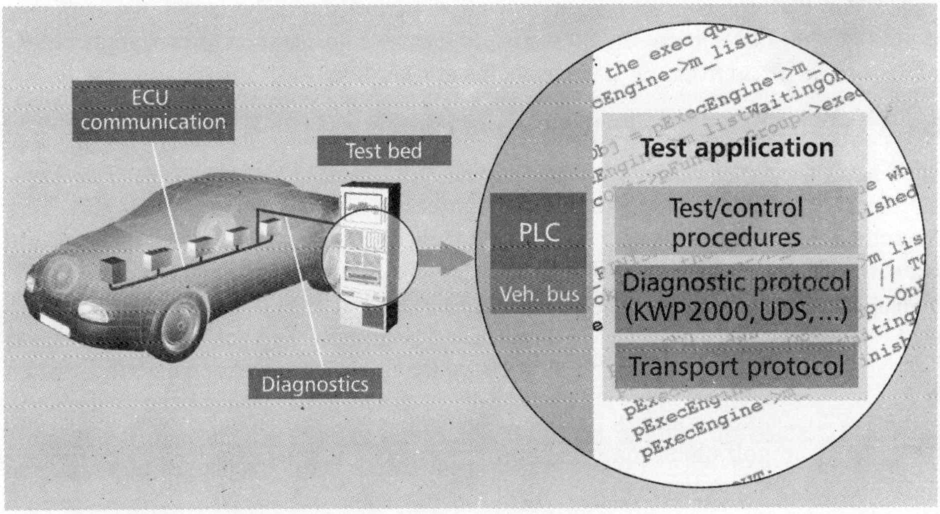

Figure 4.28: *Hardware and software of an automated test bed*

Both partners are required to ensure that the quality required is provided during the manufacturing process: the individual ECU has to be tested by the ECU supplier and the entire ECU network of the whole vehicle has to be tested by the vehicle manufacturer. In both cases, the diagnostic and transport protocols must be implemented in addition to the actual test and control procedures to realize the software for the test bed (see **Figure 4.28**).

Looking at the resulting requirements, you can see that, for example, an ECU supplier today manufactures his products taking the following conditions into consideration:

- To reduce the time and effort spent on development, a large part of the software and hardware should be able to be implemented by several customers (automobile manufacturers), even if different bus systems (CAN, FlexRay™, LIN etc.) are used with a range of diagnostic protocols (KWP 2000 (ISO 14230-3) or Unified Diagnostic Services (ISO 14229-1)).
- In the lifetime of an ECU, a large number of variants results from software extensions or during troubleshooting to which the manufacturing/testing processes should be able to be adapted flexibly.
- The completed system should be able to be tested or diagnosed as an individual component or as an integral part of the automobile.

In spite of, or rather because of, these complex requirements, the quality has to be able to be ensured as efficiently as possible. To guarantee, for example, the reusability of development services and simple adaptation to the different ECU variants, the relevant test tools must be based on an appropriate, universal standard. This standard should describe the diagnostic capabilities of an ECU (or an entire automobile) as flexibly (for all applications in question) and efficiently as possible.

And so that both the overall system, as well as part of it, can be tested, the exchange of configuration data between supplier and automobile manufacturer should be extremely easy. An appropriate standard is the answer here too.

The ASAM MCD standard – flexible and efficient

Good news for the automotive sector: this kind of standard already exists. The Association for Standardisation of Automation- and Measuring Systems (ASAM e.V. [1], a committee for standardising interfaces and exchange formats for automation and measuring systems in the automotive sector) has been working on precisely this subject for years in the form of various committees, referred to as MCD committees. MCD stands for Measurement, Calibration and Diagnosis. First of all, the format and content for the description of the protocol features and diagnostic capabilities of an ECU were standardized. This standard, called ASAM MCD-2, has been available as Version 2.0 since the middle of 2004. Since this version was issued, it has quite simply been called ASAM ODX (**O**pen **D**iagnostic Data E**x**change) although the official name is ASAM MCD-2D (ODX). Strictly speaking, the MCD-2 specification has three categories: measurement, calibration and diagnostics. This article, however, concentrates exclusively on the "world" of diagnostics.

In an MCD-2 database, the protocol features and diagnostic capabilities are described in XML format, i.e. there is a guarantee that data can be exchanged simply and long-term. An important feature of MCD-2 ECU diagnostics makes it easy for users to deal with the diagnostic data. Both the diagnostic queries and the responses from the ECU are represented symbolically. The ASAM database contains a large number of conversion tables and formulae for converting symbolic data into binary data and vice versa. The MCD-2 standard thus permits quite complex conversions.

The fundamental contents of the MCD-2 data determine the following:

- bus communication parameters, e.g. transfer rate, bit times, timeout values, block sizes
- description of the diagnostic services supported by the ECU with requests, responses, parameters, units, conversion formulae etc.
- configuration information and programming data (based on HEX/S-Rec files) for programming the ECUs

The MCD-2 data is used to configure a runtime system which carries out communication or diagnostics with the ECUs (see Fig. 4.30). The functions and the programming interface (API) of this runtime system are defined using what is referred to as the MCD-3 standard. A lot of different applications can be used with the runtime system, such as an automation application or repair shop tester GUI.

The division of the standard into MCD-2 and MCD-3 makes it possible to describe the protocol or bus parameters and the diagnostic capabilities of ECUs completely outside the runtime system in order to achieve as much flexibility as possible. When the communication characteristics of ECUs change, adaptation is easy as all you have to do is modify or exchange the description files. The runtime system itself does not have to be adapted.

This kind of runtime system, corresponding to MCD-3, is available with Softing's Diagnostic Tool Set (DTS V7) [2].

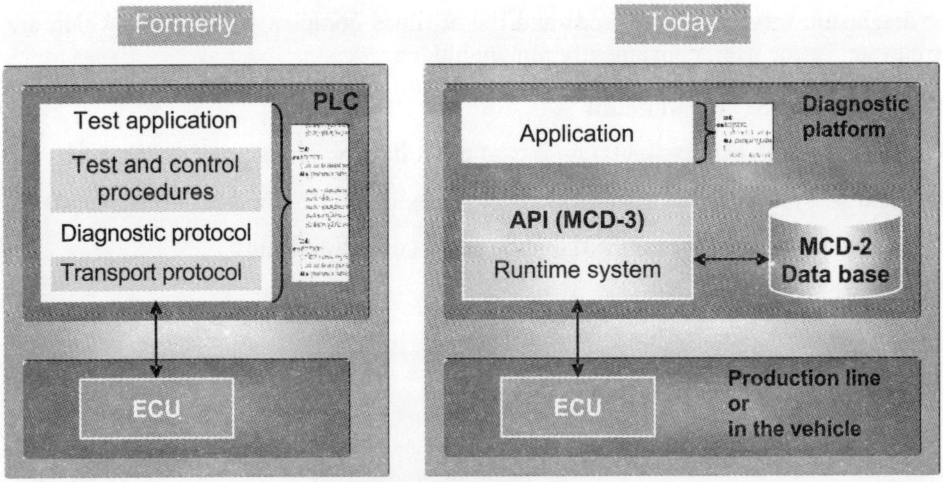

Figure 4.29: *MCD-3 runtime system with MCD-2 database*

The runtime system illustrated in **Figure 4.29**: MCD-3 runtime system with MCD-2 database contains an API which corresponds to the ASAM MCD-3 specification. The application is based on this API. The runtime system evaluates the MCD2-specific data for communication with the ECU or for its diagnostics. (It is connected to the ECU using a CAN board, for example.)

MCD-3 in the focus of manufacturing

Whereas ASAM MCD-2 focuses on ensuring simple data exchange between supplier and automobile manufacturer, the purpose of implementing a runtime system corresponding to MCD-3 in manufacturing is more about being able to ensure a synergy of engineering services already performed. This means that the time and money already invested in MCD-2 data and the MCD-3 runtime system, can be of benefit to the customer throughout the lifetime of an ECU. If this kind of standard tool is not used, there would be a constant redevelopment and adaptation of the communication and diagnostic layers.

As a result of former technical developments, most applications standard today in the area of automation (Soft PLC, SCADA etc.) still do not use an MCD-3 interface in spite of the advantages listed above. Due to its wide range of functions, the MCD-3 API, with around 1500 methods and objects, is also relatively complex. It demands extensive development skills and also a lot of time for familiarization from application developers.

Test and production systems with the DTS-OPC-Server

For years now, automation technology has been using a standard accepted worldwide – OPC technology. To be able to make use of the advantages of the ASAM standard in this area too, Softing equipped its MCD-3 runtime system DTS V7 with an OPC interface. The aim is to create a symbolic data interface for accessing automotive ECUs. This makes it simple to map data and function interfaces of the ECU to OPC data points. Details of communication (communication parameters, CAN identifiers or diagnostic services to be used) and the at times complex parameters, which are irrelevant for the user, conveniently remain hidden.

The DTS-OPC-Server in detail

The interface (see **Figure 4.30**) consists of the following elements:

1. Runtime system for communication (in acc. with the ASAM MCD-3 standard)
2. Configuration files for onboard and offboard communication
3. OPC-Server interface
4. Configuration data for the OPC interface

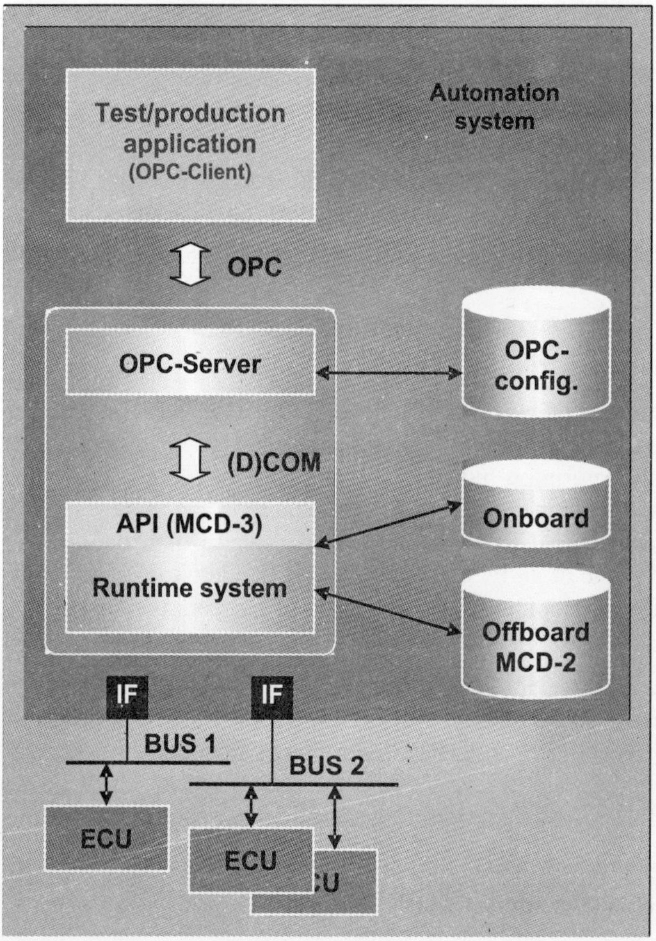

Figure 4.30: System architecture of the DTS-OPC-Server

A fundamental part of the interface is the runtime system which is responsible for onboard and offboard communication between the test system and the ECU. To configure onboard communication, the DTS-V7 runtime system can also use various proprietary description formats, such as *CANdb* (file format for describing messages in a CAN network).

The OPC-Server architecture shown in **Figure 4.30** realizes the concrete implementation of the object-oriented ASAM MCD-3 interface to a data-oriented OPC interface. As with the MCD-3 runtime system, the configuration of the OPC-Server takes place via a separate interface description file (in XML format).

The high level of abstraction becomes clear when you take a look at an – albeit simplified – description of the data flow in the DTS-OPC-Server (cf. **Figure 4.31**):

0) The assignment of the OPC variable *RPM* (rounds per minute) and of the diagnostic service *ReadDataByLocalID* (including the value for the *ID* parameter) is specified in the OPC configuration.
1) The OPC-Server initiates the diagnostic service *ReadDataByLocalID* where the parameter ID=*RPM* in the MCD-3 runtime system at a symbolic level.
2) The MCD-3 runtime system now converts this symbolic code into a corresponding binary data flow on the bus using the MCD-2 data.
3) The binary response of the ECU goes through these steps in the opposite direction.

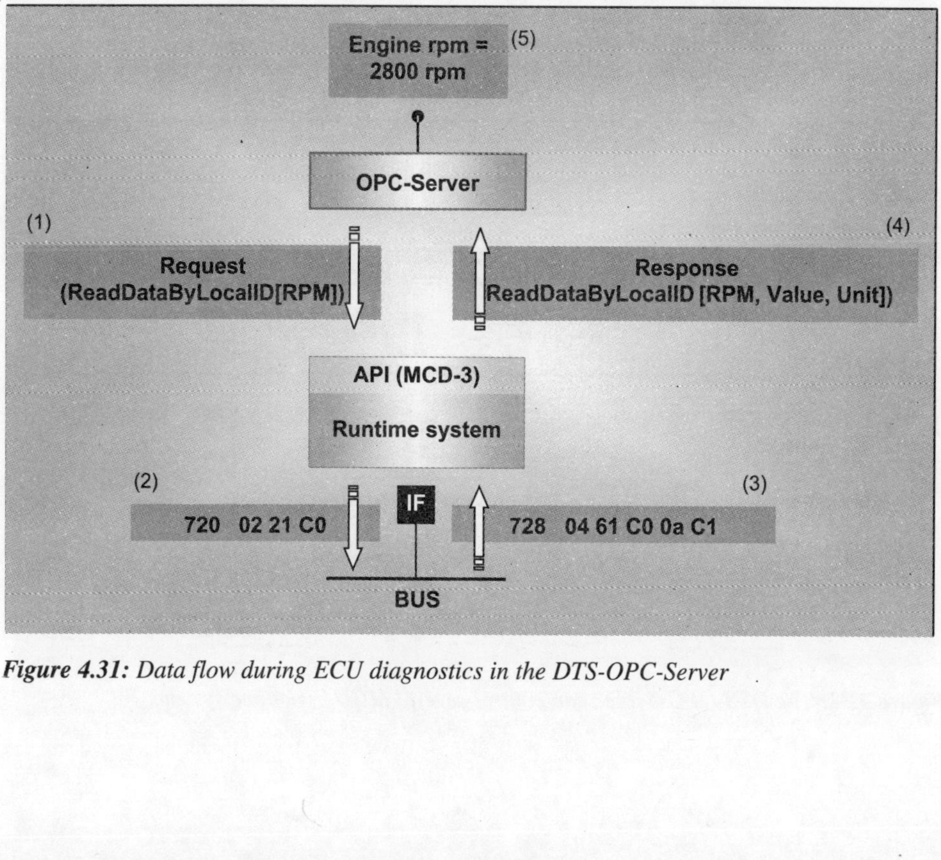

Figure 4.31: *Data flow during ECU diagnostics in the DTS-OPC-Server*

4) In the MCD-3 runtime system, the binary response is initially converted into a value in symbolic representation.

5) The OPC-Server accepts this value and sets the OPC variable to the corresponding value.

Of course it is also possible for values not only to be read by the OPC-Server but also to be written to the ECU.

If you delve deeper into the OPC name space of the DTS-OPC-Server, you will find not only the actual OPC variables for

– parameters, measurement values and target values

but also other variables which

– start processing diagnostic services,

– indicate the progress of processing in the MCD-3 runtime system and

– reflect certain error condition.

In everyday manufacturing, it is often the case that more than one ECU has to be tested. As well as the procedures described above, far more complex test scenarios can also be realized with the DTS-OPC-Server:

– parallel access of several (different) vehicle buses

– parallel access of several ECUs at one or more vehicle buses

– parallel access of several vehicles

– control of several MCD-3 runtime systems by an OPC-Server (see **Figure 4.32**)

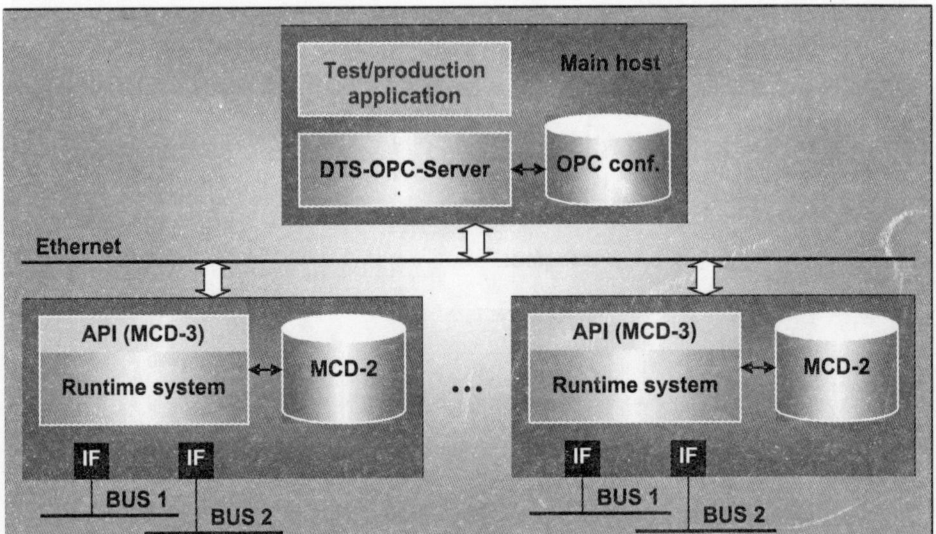

Figure 4.32: The DTS-OPC-Server can control several MCD-3 runtime systems

Configurator for OPC-Server

Which variables does the OPC-Server use in its name space? Basically, this is decided via the diagnostic services of the MCD-2 data. There is, however, a configurator (see screenshot in **Figure 4.33**) available for defining the available variables. This offers the following possibilities:

- definition and naming of the OPC variables including the link to the relevant diagnostic services
- reduction to the diagnostic services required for the test
- default values from diagnostic service parameters can be specified more precisely
- definition of which direction the data should flow in (measurement value or target value)

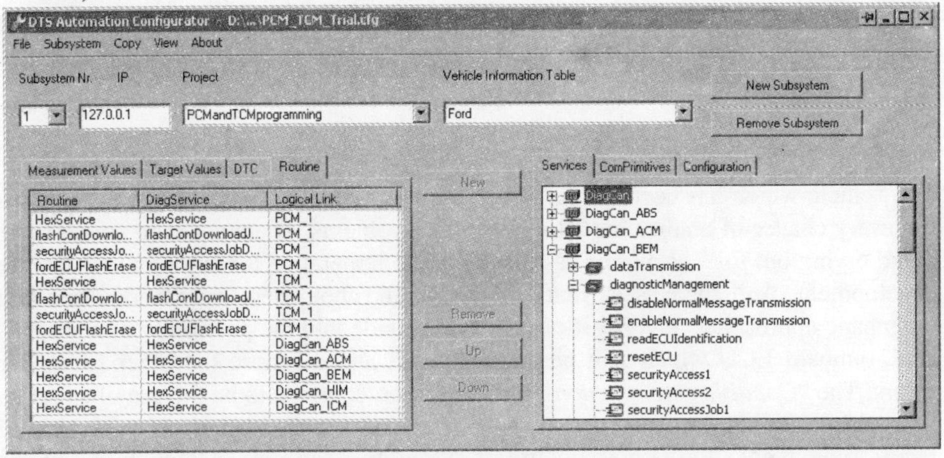

Figure 4.33: *Configurator for the DTS-OPC-Server*

The configurator (see Figure 4.33) can be used to display and select the diagnostic services ("Services"), which are available in the MCD-2 data, on the right-hand side (tree representation) as well as specify the parameters of the diagnostic services. On the left (list representation), the diagnostic services are assigned to specific categories, such as *Measurement Values* or *Target Values,* and the names of the OPC variables are displayed.

Double the benefit in automation

The method of operating the test and production system directly on the binary data of the vehicle or ECU bus is still widely used. The disadvantage of this method is, however, that the source code of the test and production systems is becoming more and more difficult to maintain due to adaptations and extensions which are necessary for the code to correspond to new requirements.

In contrast to this, using MCD-2 systems together with OPC technology in manufacturing has a number of advantages:

Advantages of OPC technology:

– Possibility of connecting to all kinds of different systems in the automation sector. These include SCADA systems (WinCC, InTouch, Citect, Genesis etc.) and PC-based controls (WinAC, 4Control etc.) as well as graphic programming interfaces (e.g. LabVIEW™)

– Simplification in comparison to MCD-3-API

Advantages of the MCD technology used in the DTS-OPC-Server:

– Possible to concentrate on the actual automation tasks. Having to deal with details of bus protocols and diagnostic services is no longer necessary.

– Changes to ECU software can be offset in the MCD-2 data. The software of the test/automation system, on the other hand, does not have to be adapted.

– Test scenarios can be standardized for all manufacturers and departments as the MCD-2 data can be used from the Engineering of an ECU through Test and Manufacturing to Service.

Dilemma with a way out

The vehicle which has been ridiculed by many a "visionary" as a *Microsoft car* thus has every chance of eradicating all its software problems. There are even voices to be heard crying out for "zero error" as a categorical imperative for automotive software development. And there is currently a lot of talk about new or extended test and diagnostic approaches. Investigations are being made into the possibility of a specialized, onboard ECU which will permanently test and diagnose the other ECUs for errors. The "*Conditio sine qua non*" will, however, continue to be the use of suitable, accepted standards. In any case, our golf enthusiast reached her goal, safely and in record time, thanks to the high quality of automotive software. Although in vain, as the accounts system in the golf club had, unfortunately, once again crashed ...

References

[1]: ASAM e.V.: http://www.asam.de

[2]: Softing AG: http://www.softing.com

Martin Dietze and *Birgit Mrowietz*, Softing AG, Munich, Germany

4.4.5 Industrial & Financial Systems (IFS) uses OPC to interact with Automation Equipment

Industrial & Financial Systems (IFS), with 3,400 employees, develops, sells, installs, and supports ERP software (for maintenance, production, manufacturing, HR, finance, distribution, etc.) in 41 countries around the world. We at IFS have developed an OPC Client that makes it possible for us to seamlessly integrate data between business applications and the industrial automation environment.

The IFS business concept is to provide business applications that allow our customers to grow, develop, and change. Our product, IFS Applications, uses proven, entirely component-based technology that makes it easy to add new e-business or ERP functionality and to integrate the solutions with other information systems. IFS is the world's fastest growing supplier of business applications, with R&D centers in six locations in Europe, Asia, and the USA.

Early in our history, IFS focused on heavy industries such as mining, energy, and forestry. Yet our system components are easily configured and adapted to meet any vertical market's needs. Today, our more than 3,000 customers are found in a wide range of industries, from Internet entrepreneurs to engineering and project delivery as well as aviation, rail, and defense industries. From our long involvement with the forest industry, we have experience of integration with industry-specific systems, such as trimming and process control.

Thanks to its flexibility and open architecture, IFS Applications can quickly adopt and employ new technologies. For example, it easily allows access via new clients such as WAP phones, PDAs, bar code scanners, and web browsers.

This openness and our early development of packaged Internet-based systems integration were the prerequisites for pioneering a crossover solution that enables a business software system to utilize technologies designed solely for automation purposes.

This integrated flow of data, which begins with an Internet-based customer order and goes all the way through production, offers many new possibilities.

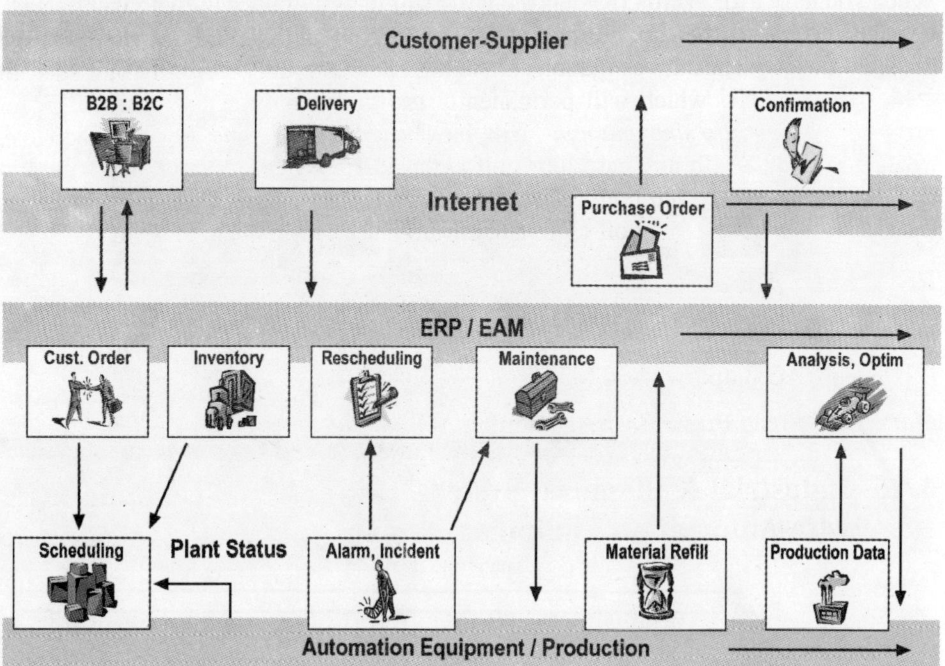

Figure 4.34: *A continuous information flow enables the total integration of the factory*

The integration brings benefits to both information environments, as critical information is exchanged both ways. The business software secures the information basis for production planning – and rescheduling at short notice – and provides tools for the analysis and optimization of availability, efficiency, and quality output. In the other direction, the data capture from the OPC can be utilized in IFS Applications for preventive maintenance, automatic generation of work orders, and warehouse replenishment processes.

We can, for example, carry out maintenance based on real-time data. An alarm on the automation side automatically creates a work order on the ERP side; a service interval is then carried out based on real-time operational hours. Users are also able to optimize the scheduling of what to manufacture – based on the actual order stock, the material on hand, and the real-time status of the plant (maybe some parts are standing still). This information is then sent directly to the automation side.

This opens a way to manufacture customized, small series, with a clean data flow all the way from the customer to the production line.

A customer can design and put his customer order in an Internet site. The customer order will be stored in IFS Applications. Together with other orders (and data from other parts of IFS Applications) a manufacturing schedule will be created and transferred to the SCADA system on the automation side. If something happens during production, a work order will be sent to repair staff so that production can be resumed as soon as possible.

When you have a good data flow between the ERP side and the automation side, there is really no end to the possibilities. Operators on the automation side are able to navigate directly from their SCADA terminals to access useful information such as

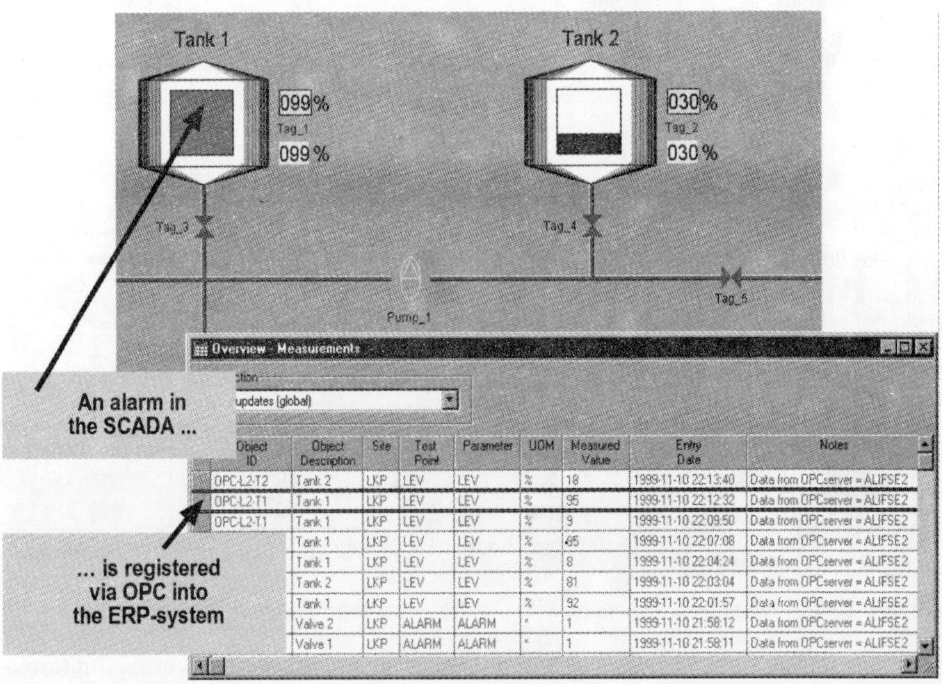

Figure 4.35: *OPC connects automation and ERP system*

history, object data, scheduled maintenance, and such statistics as fault frequencies and cause-of-error correlations. Operators are also able to create fault reports and close work orders, among other reports.

For IFS, the main focus was to find a communication standard that was widely accepted and would work with a variety of automation equipment from many different suppliers. OPC is supported by every main vendor on the automation side and uses a technology that is also familiar to ERP people. OPC was the only standard that met these demands and, for IFS, the only standard that could be chosen.

The COM/DCOM technology used in OPC matched our platform for external communication (IFS eConnect) and made it quite smooth to develop an OPC-Client adapter on our side.

Another important advantage was that we could use different SCADA systems as gateways. Most SCADA systems today act as OPC servers. This makes it possible to read/write data without having to deal with different industrial buses. We can ask for data as an OPC client (compare SCADA client) and wait until the OPC server (compare SCADA server) has brought us the data.

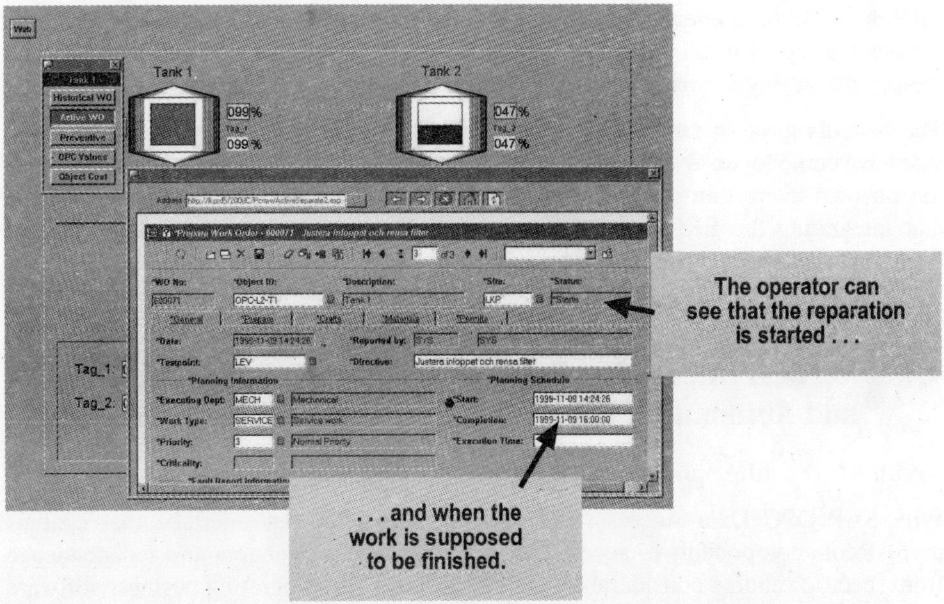

Figure 4.36: Work orders can be started and observed by the ERP system

Figure 4.36 shows how a customer order is executed via the Internet. The ERP system gets information about the order, stock status, and status of the production area. Based on this, a reasonable production schedule can be set up and transmitted directly to the production machines.

An interruption or an alarm during production generates a work order to solve the problem. During production, machine observation and other analysis and optimization systems can be supplied with data for the automation device.

There is traditionally an aversion (often justified) among staff on the automation side to allowing administrative systems into their data flow. They are afraid of interference in production. With the OPC technology we act as any SCADA client, which makes the automation people more relaxed. The intensity of the extra data flow – busload – brought in by the demands of the ERP system is in these connections (ERP – Automation) is quite low. An ERP system can in most cases sample data every 10 minutes and still find it speedy.

Furthermore, the communication between systems is secured by means of routers (and, if necessary, firewalls) that define which IP addresses are permitted access as well as limits to data load size permitted to flow between systems. This eliminates the risk of overloading the internal SCADA network.

The solution attracted a lot of interest when it was first shown in late November 1999. We showed the data flow live and we successfully managed maintenance based on incidents in the automation equipment.

Our OPC Client was developed inside IFS R&D in Gothenburg, Sweden. The OPC client is designed with 3 main setup steps. In the OPC client, IFS defines the tag name and the sample rate, and IFS defines the rules on when the data should be stored in the ERP database (e.g. every 24 hours and every time an interval is passed). IFS also defines the layout of an XML file depending on the subscriber on the ERP side and the key for storing the data. It looks a bit different in the other direction.

The benefits have so far been considerable. Three main vendors on the automation side have come forward with suggestions for co-operation, and a number of customers have shown interest in participating in projects of this kind. IFS is quite convinced that integrating the ERP and automation sides in industrial environments is set to become a significant business opportunity.

Anders Lif, Industrial & Financial Systems, Sweden

4.4.6 Vertical Integration Between E-Business Applications and Automation Using SAP OPC Data Access (SAP ODA)

Architecture and Application

With SAP OPC Data Access (SAP ODA), SAP offers an interface that enables mySAP.com components to access OPC servers. OPC's openness and independence from specific vendors considerably reduces the costs for integrating business software with the world of automation. To ensure a timely and consistent supply chain, more and more data and events are required directly from production. SAP ODA supports both the OPC Data Access and Alarms&Events specifications, which means that production data and events can be processed directly in the mySAP Supply Chain Management (mySAP SCM) solution.

The production employees of a company have, for example, direct access to individual data points in automation from the SAP process instruction sheet (PI sheet). This means that entering data twice, which increases the likelihood of errors, is no longer necessary. In addition, the SAP PI sheet can react to OPC value changes and events

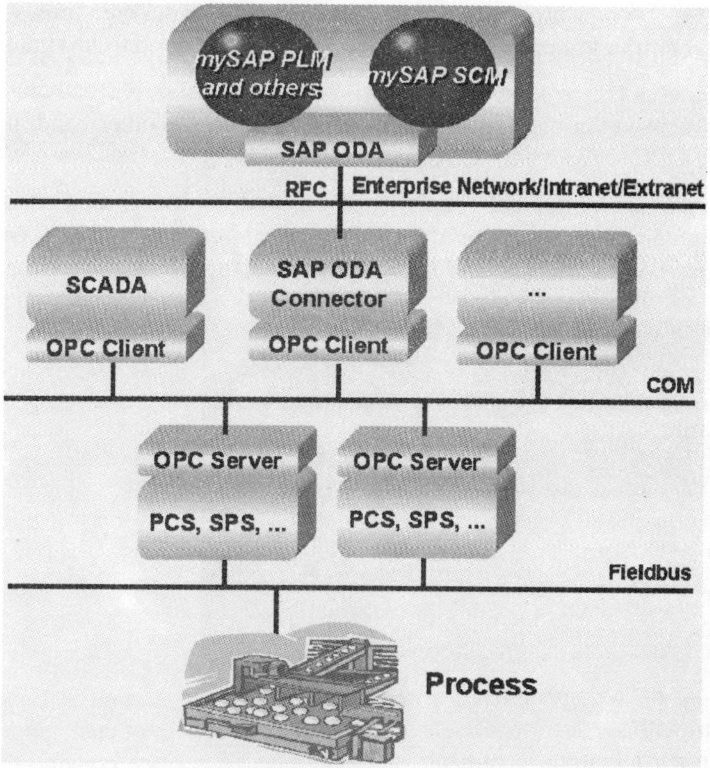

***Figure 4.37:** Architecture for Vertical Integration Using OPC*

and thus provide the user with situation-related support. Follow-up activities, such as goods movements, confirmations of order statuses or maintenance tasks, are automatically initiated in the system through process messages. All activities and parameters can be logged and archived in an electronic batch record (EBR).

OPC data points (items) can also be queried periodically or if a specific event occurs. This ensures a complete documentation of the production process.

As a generic interface, SAP ODA can, in principle, be used to supply any mySAP.com component, such as SAP BW or mySAP SCM Supply Chain Event Management, with production data.

Technology

Technically speaking, the SAP ODA interface consists of two parts – the programming interface (API) within the SAP R/3 System and the SAP ODA Connector outside of the SAP R/3 System.

The system-internal programming interface is written in ABAP and can be used by any SAP component. In addition, customer-specific event processors can be called via Business Add-Ins (BAdIs).

No extra programming is required to define logical names for OPC data points and events, that stand for the technical settings for the Remote Function Call (RFC) and

OPC communication. OPC server browsing, namespace browsing, event area browsing, and the query of event filters supported by a server all facilitate the configuration.

The SAP ODA Connector acts as a kind of gateway – on the one hand, it communicates with the SAP System(s) via the network using RFC, and on the other hand, it communicates with the OPC servers via COM.

The SAP ODA Connector's operational reliability is ensured by installation as a Windows service, an automatic reconnection if the OPC server shuts down or RFC or network problems occur, event buffering, and recovery mechanisms.

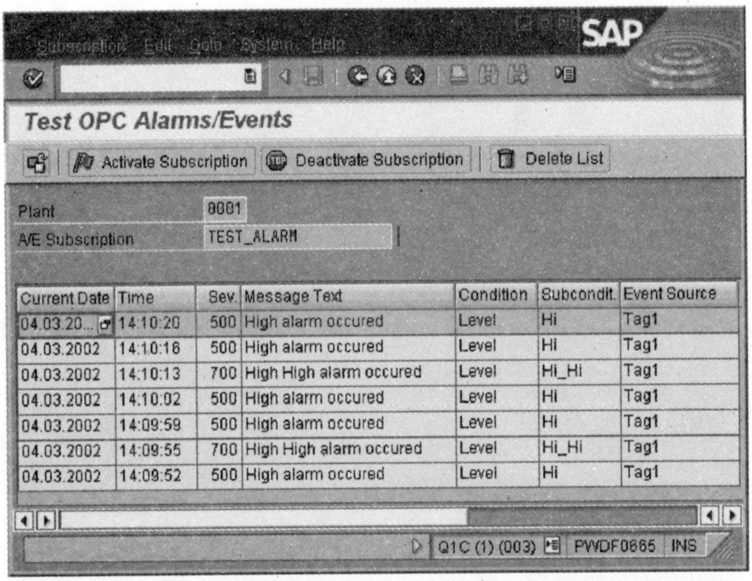

Figure 4.38: *OPC Events in SAP R/3*

As the SAP Web Application Server also supports XML and SOAP, it will in future be possible to access OPC servers via the http protocol. On the other hand, it would also be feasible to read the definition of logical data points as XML from the SAP System and thus support the automatic configuration of OPC servers.

SAP ODA 2.0 supports OPC specifications mentioned in **Table 4.1**.

Table 4.1: *Functionality of SAP ODA 2.0*

OPC Specification	Individual Functions Supported
Data Access 1.x and 2.x	Server Browsing Namespace Browsing Synchronous Read and Write Operations Subscription to Value Changes
Alarms and Events 1.x	Event Area Browsing Query of Available A/E Filters Subscription to Alarms/Events

SAP ODA 2.0 is part of the standard shipment of SAP R/3 Enterprise. No additional license fees are charged. In addition, SAP customers can download the software from the SAP Service Marketplace (http://service.sap.com) free of charge.

Example: Integration Between the mySAP SCM Solution and the Process Control Level in the Food and Process Industries

Aarhus Olie ranks among the largest producers of vegetable oil and speciality fats for the food industries. In January 2002, they went live with the mySAP SCM solution in their plant in Denmark. One of the components they use is the PP-PI component for process industries. The new solution replaces old mainframe-based systems.

In the production environment, Aarhus Olie set a number of additional goals in connection with the implementation project:

- A more efficient, system-aided coordination between the planner and process operator as well as between different production departments
- Better production traceability and visibility for a faster and more flexible reaction to customer demands
- Low maintenance and implementation costs for the standard software

Aarhus Olie uses several different SCADA and process control systems, such as ABB SattLine, SIMATIC PCS7 by Siemens, and systems by Fisher Rosemount, to control

Figure 4.39: *View of Tank Farm at Aarhus Olie*

and monitor their semicontinuous and batch production processes as well as the tank farm of 500 tanks.

To achieve the project goals, the existing control systems had to be integrated as best as possible, in other words, as synchronously as necessary, openly, scalably and at the lowest possible costs. For this reason, Aarhus Olie chose OPC as an open industry standard. With SAP OPC Data Access, no further middleware or vendor-specific programming was necessary to link the systems.

Using Data Access namespace browsing, more than 1,500 Data Access data points in the system have been put into a business context so far. They are defined as logical data points in the SAP R/3 System and can be used in master recipes or process orders.

In the browser-based PI sheet of the SAP solution, it is possible to transfer order- and product-related data and parameters, such as the process temperature, material number, or quantity to be produced, directly to the control system. Values such as actual quantities, work performed, machine runtimes, and so on, can be read in the PI sheet using OPC. The daily print-outs of shop floor papers becomes superfluous. Changes made to the planning, which up until now required coordination by telephone, can now be communicated much faster and more flexibly.

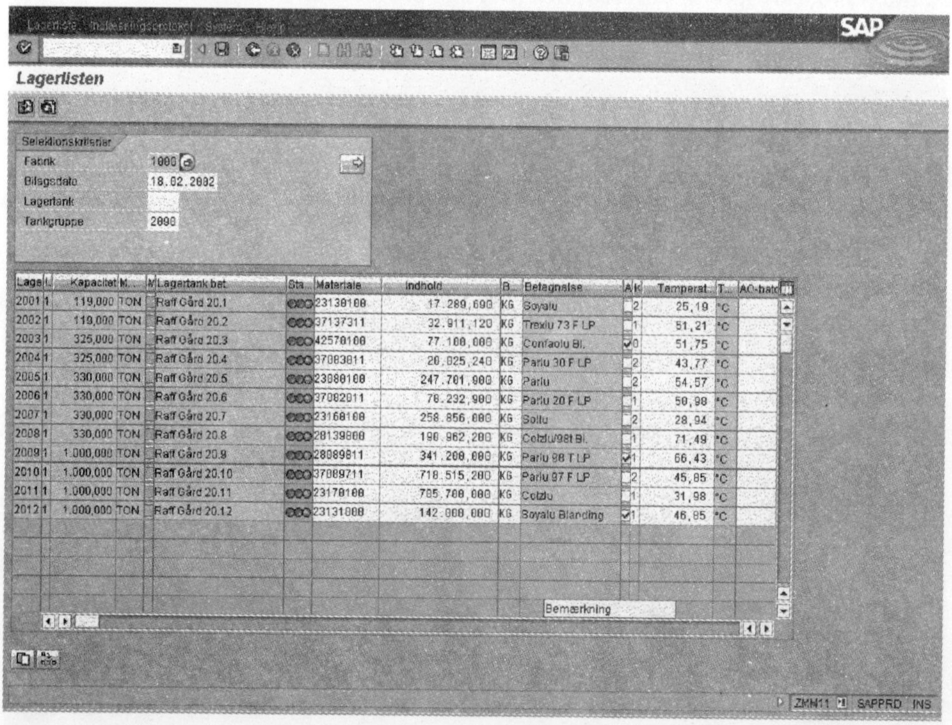

Figure 4.40: *Current Status and Stock Overview by Tank Group with SAP ODA in the mySAP.com Application at Aarhus Olie*

Since order-specific data is retained in the OPC server, this also means that the process control level can work self-sufficiently for a certain period of time.

Furthermore, a current stock overview of the tanks was realized in the customer project. Here, material numbers, quantities, temperature, quality, and tank status (for example, pump on) are read synchronously from the different OPC Data Access servers.

The project was realized in 9 months.

Aarhus Olie plans to use SAP ODA version 2.0 to benefit from Alarms&Events and the subscription to value changes.

Example: Total Automation of Crankshaft Manufacturing Using SAP ODA

Albert Weber GmbH is an established specialist for the manufacture of sophisticated components and systems for the vehicle industry. Among their customers are many of the biggest and most important automotive manufacturers.

Albert Weber GmbH also uses the mySAP SCM solution. One of the components is PP.

The goal was to automate the entire crankshaft manufacturing that is carried out in three-shifts. To achieve this, the SAP R/3 System had to be closely linked with automation systems – in this case, with seven OPC servers of the individual cell control systems. In addition to the OPC Data Access standard, Alarms&Events was

Figure 4.41: Fully-Automated Crankshaft Manufacturing at Albert Weber GmbH Markdorf with Direct Link Between SAP R/3 and the Cell Control System Using OPC

used for the first time. At present, more than 3,000 data points are used in the SAP R/3 System.

The current status of the line, the individual manufacturing cells, their machines, and the emergency system are displayed in the enterprise software. Sources and destinations of pallet transports at the line's primary loop as well as the supply of the cells with parts via the relevant carrier is determined by the SAP R/3 System. Any manual supply of the machines with crankshafts is also reported to the SAP R/3 System.

Confirmations are made for every single part along with the serial number and pallet.

With this project, Albert Weber GmbH succeeded in automating their crankshaft manufacturing completely and continuously, without implementing any further middleware – an example for Collaborative Production Management becoming real.

Summary

As a vendor-independent standard, OPC can make an important contribution to the vertical integration between the world of automation and e-business applications. The SAP OPC Data Access (SAP ODA) provides SAP customers with an integration to OPC servers.

Users of the mySAP.com e-business platform thus now have a tool for reducing and restricting the costs for integration.

It helps to make data and events from production and automation more visible for the entire supply chain and allows enterprises to better react to exceptional situations.

Using SAP ODA with the mySAP SCM solution (for example, the browser-based PI sheet), reduces the likelihood of errors due to double data entries or inconsistent data.

Holger Kunitz and *Martin Schmidt*, SAP AG, Walldorf, Germany

4.4.7 OPC and 4CONTROL Enhance Paper Quality

The Stora Enso Company

Stora Enso is a leading international company in the forestry industry. In the Paper, Packaging and Forest Products divisions, Stora Enso produces publication and fine papers, packaging boards and wood products – areas in which Stora Enso is a global market leader. In 2003, Stora Enso achieved a turnover of 12.2 billion EUR. The company has approx. 44,000 employees in more than 40 countries in five continents and reaches a production capacity of 15.7 million tons of paper and board as well as 7.4 million cubic meters of sawn wood products.

A New Product Requires New Solutions

The Reisholz plant produces uncoated magazine papers for the Publication Paper unit. In the wake of launching a new product with enhanced quality, the existing facilities, in particular the calenders, were provided with new instrumentation and control technology.

Figure 4.42: Stora Enso plant in Reisholz, Germany

Decision for OPC

To improve the operating reliability and serviceability of the facilities as well as to ensure consistency in the communication between all I&C systems, the open communication standard OPC was chosen as a common data interface. OPC is used for both communication between the individual I&C systems and communication with the Mill Information System (MI system). An OPC interface had been demanded by all suppliers of I&C technology already when the contract had been awarded.

OPC, the Open Communication Standard

The use of OPC allows a direct data exchange between the I&C components from different manufacturers, without gateways and without customizing the individual bus protocols.

The use of OPC has the following advantages:

- OPC is a defined, neutral and open data interface
- OPC ensures a smooth and neutral exchange of data between systems from different manufacturers
- OPC allows standardizing the communication
- OPC facilitates troubleshooting

Clear Rules for the Production Data Network

The Reisholz plant places great importance on a consistent use of standards and safe technologies to ensure reliable and troublefree production, good serviceability of the facilities and fast fault correction.

For these reasons, Profibus, OPC, Ethernet and fiber optics are used in the production automation.

The proven Profibus is increasingly employed for communication between the field and automation levels.

To facilitate communication between the automation components and the supervisory control system, Stora Enso basically uses OPC for its plant in Reisholz. Communication between the newly implemented I&C equipment is also performed via OPC.

The automation stations are networked via Ethernet, which is also used for communication between the automation and management levels. A 100-Mbit network is used throughout the plant.

Wherever communication lines extend beyond a room, fiber optic cables are used as the communication medium due to their high noise immunity.

4CONTROL, the Process Data Manager

To improve the handling of the provided process data, Stora Enso decided to use the 4CONTROL automation system from Softing AG as an intelligent process data manager between the MI system and the I&C systems.

The systems used have either OPC server or OPC client functionality. Therefore, not all the systems can communicate with the MI system, which has only an OPC client. This problem could be solved with 4CONTROL, which features both an integrated OPC server and an OPC client by default.

4CONTROL, the Data Integrator

4CONTROL communicates with the following systems:

- Airtech from V.I.B.
- Glosstech from V.I.B.
- DaVinci from Honeywell
- Damatic from Metso
- DeltaV from Emerson
- MOPS from Tieto Enator

Figure 4.43: *The calender has a decisive influence on the quality of the paper*

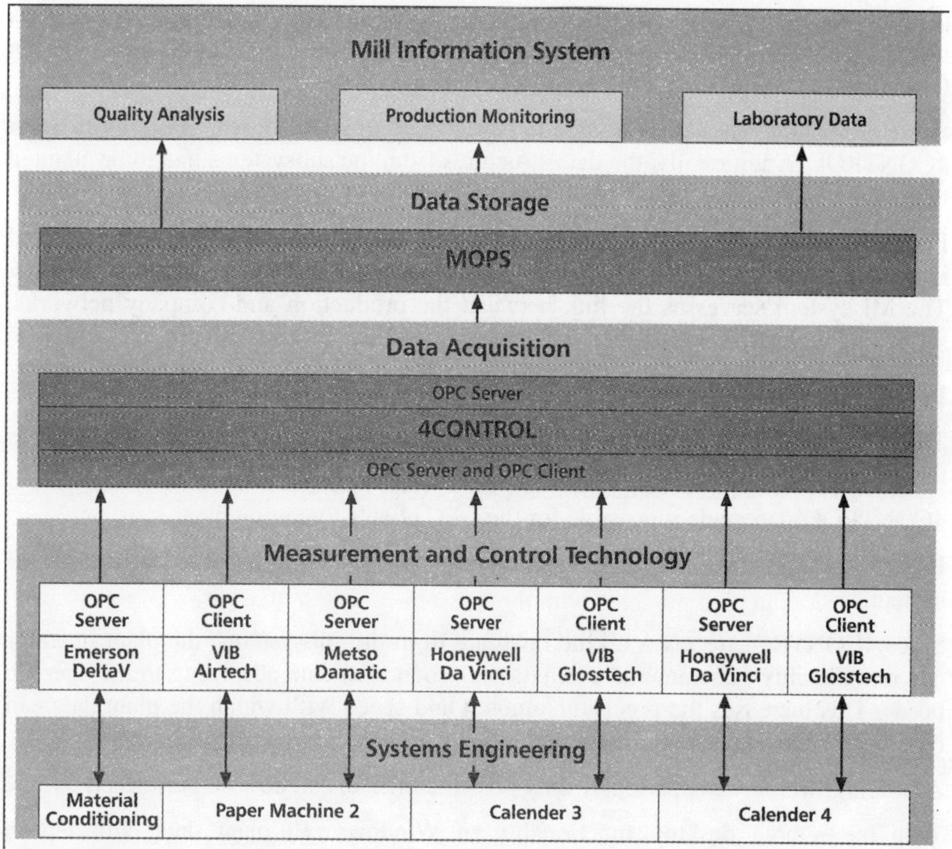

Figure 4.44: *Process data management with 4CONTROL and OPC*

4CONTROL reads the data from these systems over its OPC client and OPC server interfaces, and forwards the data to the MI system (OPC client) over the OPC server interface.

4CONTROL Allows Smooth OPC Communication

Part of the data supplied by the I&C systems require preprocessing before the MI system can process the data. To eliminate the need to adapt the lower-level and higher-level systems to each other, the data are adapted between the I&C systems and the MI system. This is done in 4CONTROL, which interconnects these systems. The OPC server and OPC client interfaces of 4CONTROL as well as its support of all IEC 61131-3 programming languages allows supplying the MI system with the data in the required format.

Before the use of 4CONTROL, the communication between two systems that have only an OPC client interface was not possible. To enable them to exchange data, software with OPC server functionality and internal processes for handling the interfaces is essential – 4CONTROL. In addition, the data from the subsystems needed to be converted. A small program in 4CONTROL, which preprocesses the data, ensures that only the relevant data are passed on to the MI system in the required format. This reduces the data in the database system right from the start, as redundant data are not forwarded to the database, thus avoiding a data graveyard.

Some of the OPC tag names needed to be renamed or adapted, which is easy to do in 4CONTROL. Additionally, the data structures from the subsystems had to be adapted to the quality database, which is also an easy job for 4CONTROL.

In the MI system, all the plant data (approx. 15,000 online data points) are collected in an Oracle database. The data are evaluated and analyzed by Windows client PCs. The MI system serves as the link between the production and company networks, separating them from each other.

Benefits for the customer

The use of 4CONTROL substantially reduced the system integration costs as well as the time required for integrating the data of the different systems.

4CONTROL also made it possible for the OPC clients to communicate.

In addition, a central OPC data collection point was to be implemented before the MI system.

Since 4CONTROL receives all the raw data from the subsystems, the plant operator can now flexibly generate and evaluate the production data according to the specific needs. This increases the precision, amount and speed with which the plant data can be processed, evaluated and presented.

Averaging, which was previously only possible offline, can now be performed online.

With the remote desktop functionality of Windows, all plant data collected in 4CONTROL can also be evaluated remotely. Faults are thus easier to remedy, and changes are safe and easy to manage.

Figure 4.45: *Paper quality under constant control*

In addition, the different data formats can now be flexibly adapted.

With 4CONTROL, the plant data are processed and evaluated independently of the plant's I&C systems.

Data combinations that used to be impossible or needed to be done by the vendor's service staff are now easily performed.

None of the systems needs to be modified to allow a more flexible data evaluation; the raw data are compiled to new data in 4CONTROL exactly as needed by the MI system. The existing I&C components do not have to be adapted for this purpose since 4CONTROL takes care of that, too.

The programs in 4CONTROL are created according to the IEC 61131-3 programming standard, which is easy to understand. Special knowledge about programming the different systems thus is not necessary.

The Upshot: 4CONTROL and OPC, an Excellent Team

4CONTROL and OPC not only contributed to a more open and flexible communication between the plant control and the MI system as well as to a demand-oriented data processing and evaluation. They also helped reduce the costs for investments and system integration compared to alternative solutions.

Gero Kura, Softing AG, Munich, Germany

4.5 Further aspects of using OPC products

4.5.1 Performance assessments of DCOM applications

Section 1.3 stated that OPC products provide very good performance without furnishing proof of this statement. This section establishes this proof for *Data Access Servers* and *Data Access Clients*.

The setup for the performance tests conducted consisted of the following PCs:

- 2 Pentium II PCs with 200 MHz
- 1 Pentium II PC with 400 MHz
- 1 Pentium II PC with 550 MHz

Apart from the Windows NT operating system, the *Data Access Client* and *Data Access Server* were the only applications running on the PCs. The measurements were taken for the following constellations:

- Client and server are running on one PC (200 MHz, 400 MHz, 550 MHz).
- Client and server are running on two different PCs (200 MHz). In this case, the PCs were connected via a 10Mbit/s Ethernet network. No other PCs were connected to the network except the test PCs. This ensured that no additional data traffic occurred in the network.

The following factors were varied for the individual measurements:

- The number of OPCItem objects. The number was changed from 0 to 20,000 in increments of 400. All OPCItem objects were of the same data type, i.e. Boolean. The values of the OPCItem objects were read from the PC memory by the Data Access Server. Their values were also changed in the PC memory by simulation. This means that the time for acquiring the data was only taken into account in the measurements as far as reading from the memory was concerned. Peripherals were not accessed. As all values were of the Boolean data type, the simulation consisted in a permanent change between "TRUE" and "FALSE."
- The number of OPCGroup objects. The OPCItem objects were divided up among 1, 10, 50, and 100 OPCGroup objects. The PercentDeadband parameter was set to the value 0 for the created OPCGroup objects; this means that a value will be transferred whenever a change occurs.
- The rate of change for the OPCItem objects. In real applications, the values of all OPCItem objects do not change at the same time. Measurements were taken for 0%, 5%, 10%, 50%, and 100% of the values changing simultaneously.

The measurements were conducted in the following way:

From time 0, the *Data Access Server* reads the values of all OPCItem objects that have been created for this measurement, for all OPCGroup objects. This is followed by the calculation which values have to be passed to the client. In the next step, the values to be transferred are passed to the client's callback interface by calling OnDataChange. The client does not display the values but discards them. The call therefore immediately returns to the server. On termination of the call, the time that has elapsed is measured. This sequence was repeated several times. This served to eliminate influences on the measuring result which may arise from the properties of NT (preemptive multitasking with limited assignment of priorities). The measurements thus only applied to the unrequested transmission of data from the server to the client.

The aim of the measurements was to find the minimum achievable update rate depending on the variables. Tests were conducted for the following questions regarding the use of *Data Access Servers*:

- Which update rate can be achieved for different numbers of OPCItem objects? The value can be used for determining the response time of the Data Access Server.
- What effect does allocating the OPCItem objects to a varying number of OPCGroup objects have on the minimum update rate? The transmission of a large number of OPCItem objects at once is efficient but prevents the optimum adjustment of the server to the application, as the values for PercentDeadband and UpdateRate are the same for all OPCItem objects. The result of these measurements also allows the drawing of conclusions on the allocation of the OPCItem objects to OPCGroup objects.
- How are these values influenced by distributing the server and client to two PCs? The evaluation of this result allows an assessment of the disadvantages for the performance caused by this distribution.

– Which effects do the different rates of change of the OPCItem values have?

– How does the minimum update rate change if the PC's clock frequency is increased? Does a doubling of the clock frequency result in half the update rate?

The components used for the measurements were created by using the C++ toolkit introduced in section 3.3.2.

The test results are represented in the charts below.

The chart in **Figure 4.46** shows the change in the average update rate depending on the number of OPCItem objects for a varying number of OPCGroup objects. The client and server are on the same PC.

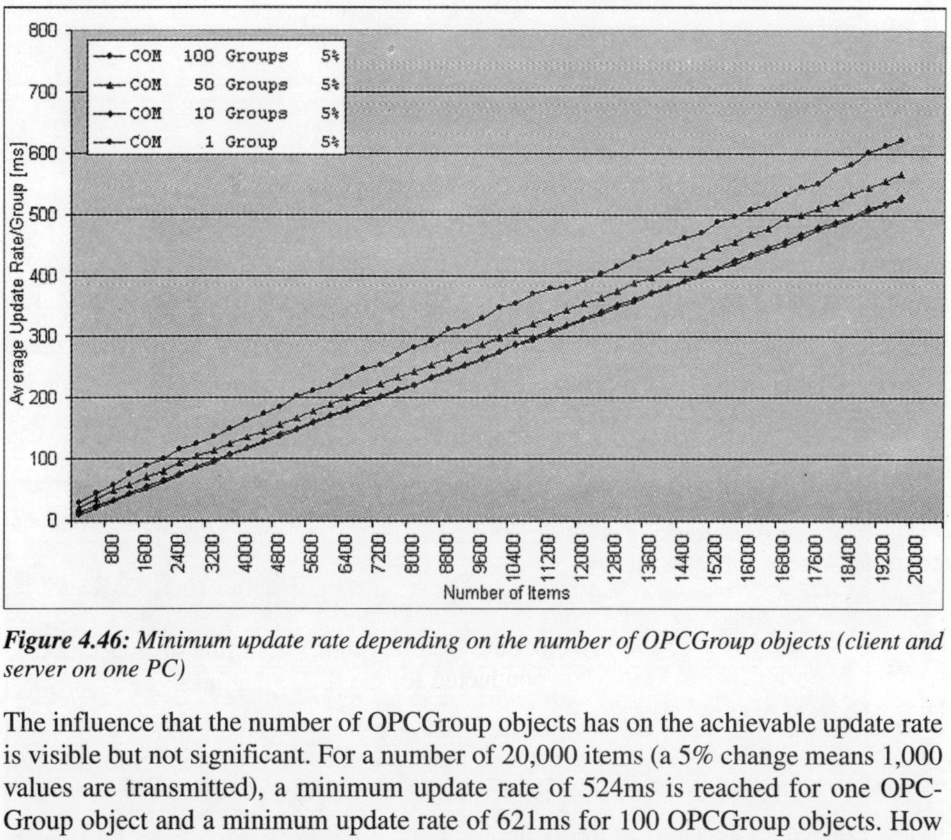

***Figure 4.46:** Minimum update rate depending on the number of OPCGroup objects (client and server on one PC)*

The influence that the number of OPCGroup objects has on the achievable update rate is visible but not significant. For a number of 20,000 items (a 5% change means 1,000 values are transmitted), a minimum update rate of 524ms is reached for one OPC-Group object and a minimum update rate of 621ms for 100 OPCGroup objects. How small the difference is for a varying number of OPCGroup objects under the described marginal conditions also becomes clear from the nearly identical courses of the lines for 1 and 10 OPCGroup objects. With a small number of OPCItem objects, the difference between the minimum achievable update rates is even less. The reason is that the allocation of OPCItem objects to OPCGroup objects is independent of the rate of change. This means that not 5% of the OPCItem objects in an OPCGroup object change at the same time but 5% of all OPCItem objects. With a distribution to 100 OPCGroup objects, it may well occur that no value changes for an OPCGroup object and thus nothing needs to be transmitted.

The chart in **Figure 4.47** shows the change in the minimum update rate if the client and server are located on different PCs. The other marginal conditions are the same as those of the chart in **Figure 4.46**.

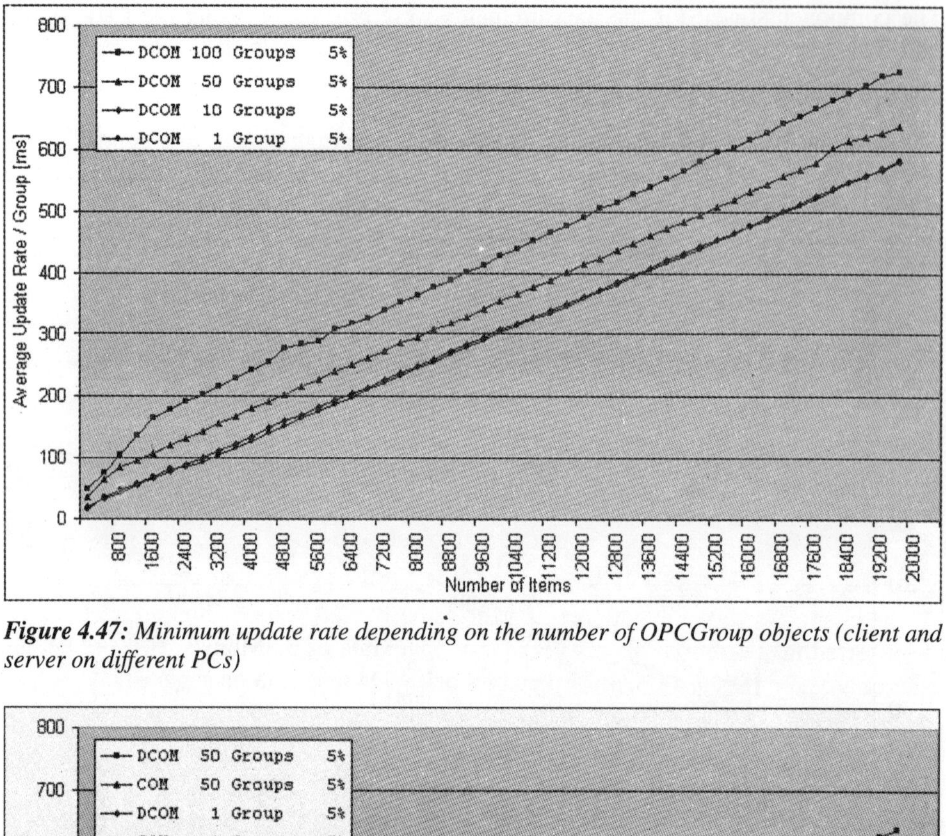

Figure 4.47: *Minimum update rate depending on the number of OPCGroup objects (client and server on different PCs)*

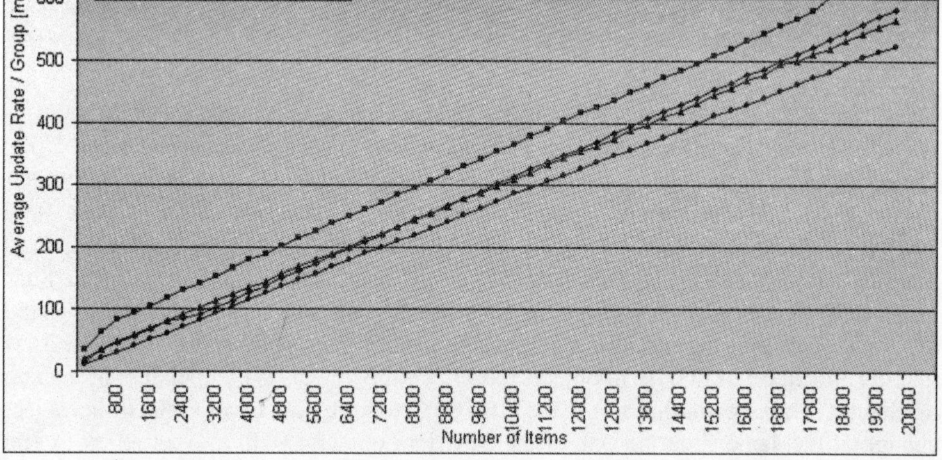

Figure 4.48: *Comparison of the minimum update rates depending on the number of OPC-Group objects for different configurations*

The difference for the achievable update rate has increased in the case of a varying number of OPCGroup objects, compared to running the client and server on the same PC. How great the difference actually is can be seen from the chart in **Figure 4.48**. The values for 1 and 50 OPCGroup objects differ clearly. When looking at the chart, it becomes obvious that using the client and server on one PC results in lower minimum update rates than running the two components on different PCs. This was to be expected. The difference is not significant, though. The value for the minimum update rate for 20,000 OPCItem objects and a 5% change for 1 OPCGroup object is 524 in the first case and 584 in the second case and, for 50 OPCGroup objects, 566 in the first case and 638 in the second case. Using the network apparently is of little consequence.

The measurements represented so far have all been taken with a rate of change of 5% for the OPCItem objects. What values for the minimum update rate can be achieved with other rates of change? The charts in **Figure 4.49** and **4.50** reflect the minimum update rates for the constellations of client and server on one PC and client and server on two PCs.

The number of OPCGroup objects of 10 is constant for both measurements. Two important results can be read from the charts. First, this is the minimum update rate for a change of all OPCItem objects. Here, the net data rate is to be calculated. The calculations refer to all OPC values that are transferred to the client. **Table 4.2** lists the parameters and their sizes in bytes.

If the client and server are running on the same PC, 20,000 OPCItem objects can be transferred in 1,008ms. This results in a net data rate of 3.8Mbits/s. If the client and server are on different PCs, a net data rate of 1.67Mbits/s can be achieved.

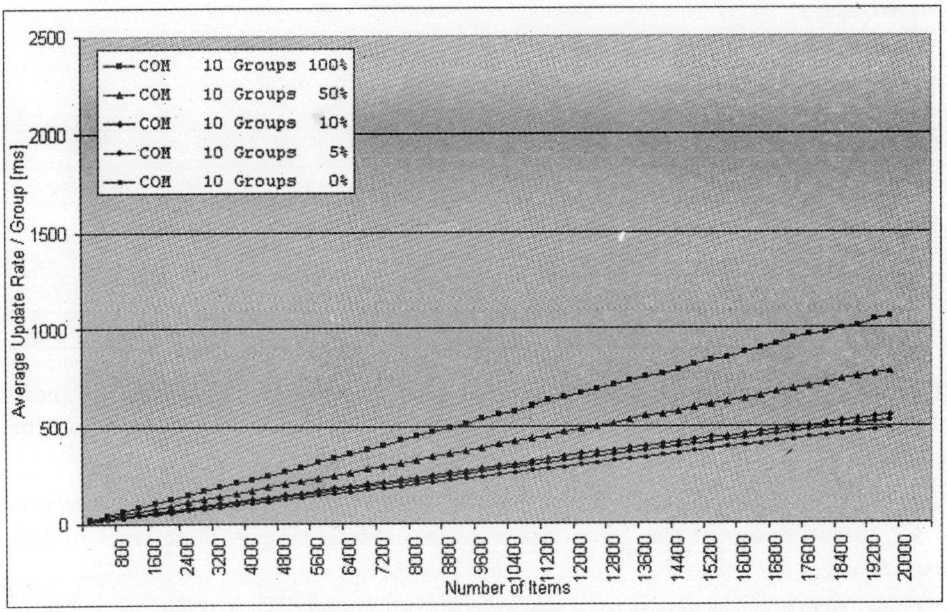

Figure 4.49: *Minimum update rate depending on different rates of change of the OPCItem object values (client and server on one PC)*

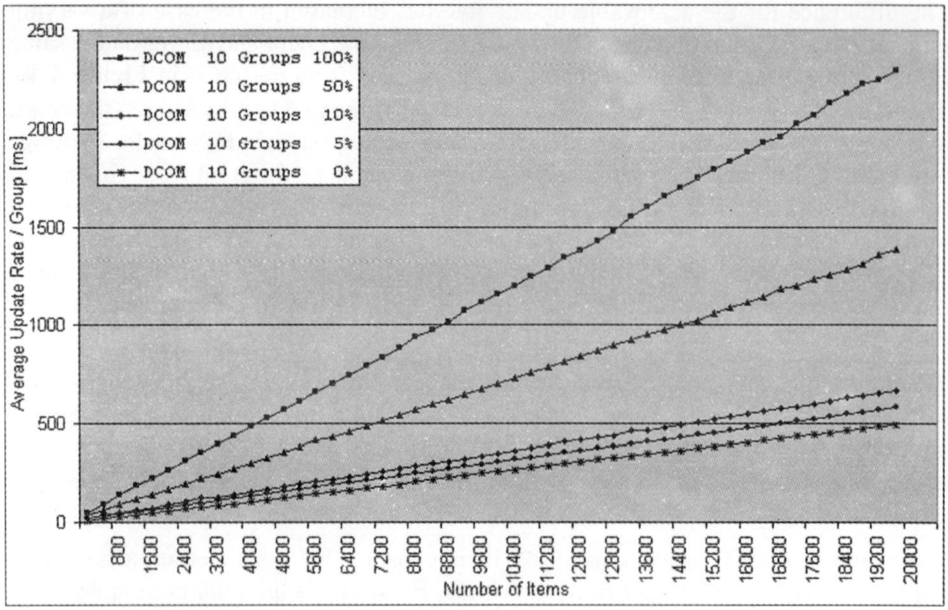

Figure 4.50: *Minimum update rate depending on different rates of change of the OPCItem object values (client and server on different PCs)*

Table 4.2: *Data quantity in the transmission from the server to the client*

TransactionIdentifier	4 byte
GroupHandle	4 byte
MasterQuality	4 byte
MasterError	4 byte
Number of OPCItem objects in the call per OPCItem object	4 byte
ClientHandle	4 byte
Value	6 byte (4 byte value + 2 byte type information)
Quality	2 byte
TimeStamp	8 byte
Error	4 byte
All together	480,200 byte for 20,000 OPCItem objects

The other important result that can be read from the charts is the internal overhead that occurs for reading out the values from the memory and for comparison. The course of the line for a 0% change corresponds to this behavior.

The value of this overhead is specific for the setup of the conducted tests. In section 3.3.2, it was explained that *Data Access Servers* should support three types of data exchange:

– Direct I/O to access data that exist in the PC.

– Queued I/O to request data from devices and receive them asynchronously.

– Event-driven I/O to receive data that are transmitted by devices without having been requested.

For the different types of data exchange, different values will result for the overhead. In the case of event-driven I/O, the comparison is nearly the only effort. The values only need to be received. A value that is not received does not have to (and cannot) be checked. With queued I/O, all values need to be requested and compared. As the response does not arrive immediately, the corresponding effort for receiving the responses will also occur. Although all values also need to be read and compared in the case of a synchronous I/O, it is not necessary to distinguish between requesting and receiving the values.

The next chart shows the direct comparison between the measurements if the two applications are running on one or on two PCs.

What do the results presented so far mean and what can be deduced?

All OPCItem objects were of the same VARIANT data type, i.e. Boolean. Boolean is mapped on 4 bytes in DCOM. The charts can thus be applied to all VARIANT data types that are also mapped on 4 bytes.

The measured values can be used for design decisions. They can provide answers, for example, for practical questions of the following kind:

A plant operator needs to be informed about a change in the process within 0.5 seconds. How many OPCItem objects may exist in the *Data Access Server*, on the assumption that 5% of the values change at the same time? The overall consideration also needs to take into account the values for data acquisition in the process and the visualization of the value.

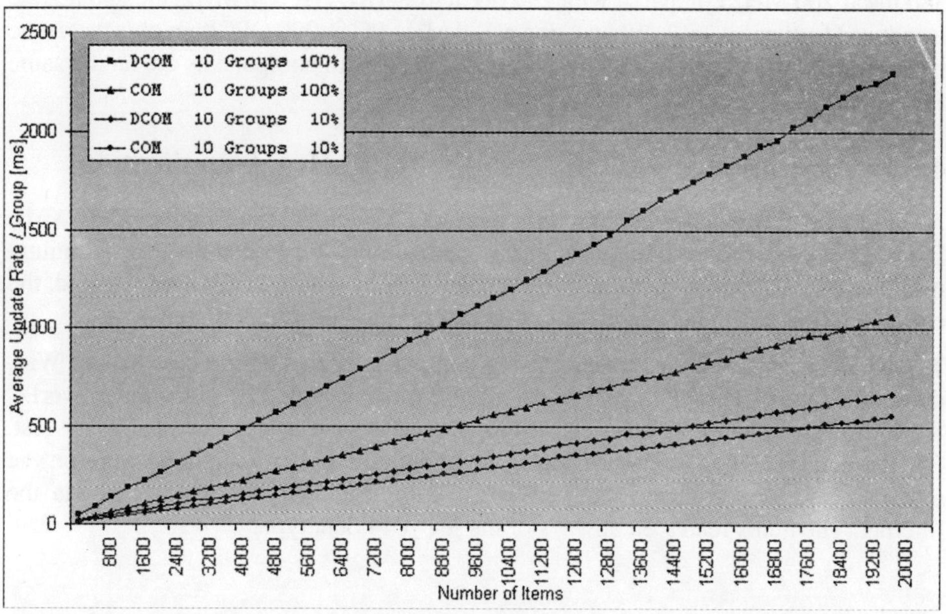

Figure 4.51: *Minimum update rate depending on different rates of change of the OPCItem object values for different configurations*

The chart in **Figure 4.52** shows the minimum achievable update rate for different numbers of OPCItem objects depending on the processor speed. The OPCItem objects are always divided into 50 OPCGroup objects.

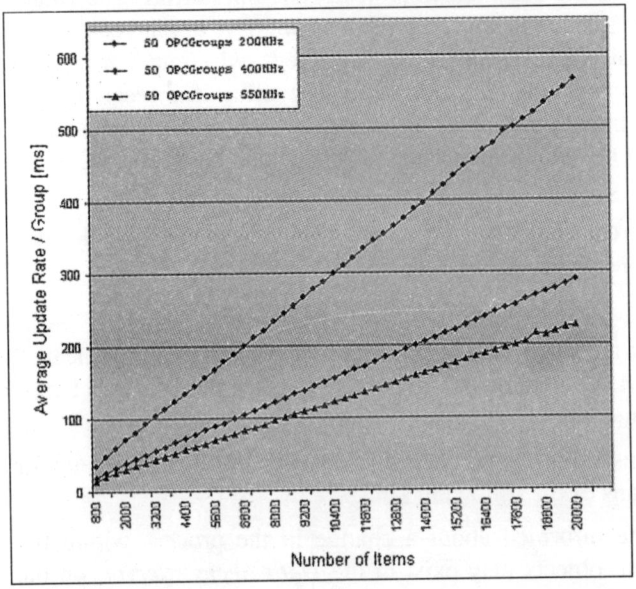

Figure 4.52: *Minimum update rate depending on the processor speed*

It may be surprising – a double clock frequency leads to nearly half the update rate – 580 ms at 200 MHz compared with 299 ms at 400 MHz. At 550 MHz, the update rate is reduced further by 25%. When using a 550 MHz PC, 1,000 OPCItem objects can be transferred from the server to the client in 227 ms if both components are on the same PC. There is thus a point in using a faster PC.

4.5.2 Performance assessments of OPC XML DA applications

As described above when dealing with the OPC XML DA specification, OPC XML DA applications do not support callback connections. Consequently, the minimum response time cannot be measured in the same way as with OPC DCOM. Instead, the entire duration of a synchronous read was determined for different scenarios.

The measurements were conducted using computers with Intel processors and Windows and Linux operating systems. In addition, one computer featuring a PowerPC 855T (Motorola) with 50 MHz and Linux operating system was included in the test. For the majority of the measurements, the client and server computers were on the intranet. One measurement was performed with the computers connected via the Internet. The items read were always of the VT_UI4 data type.

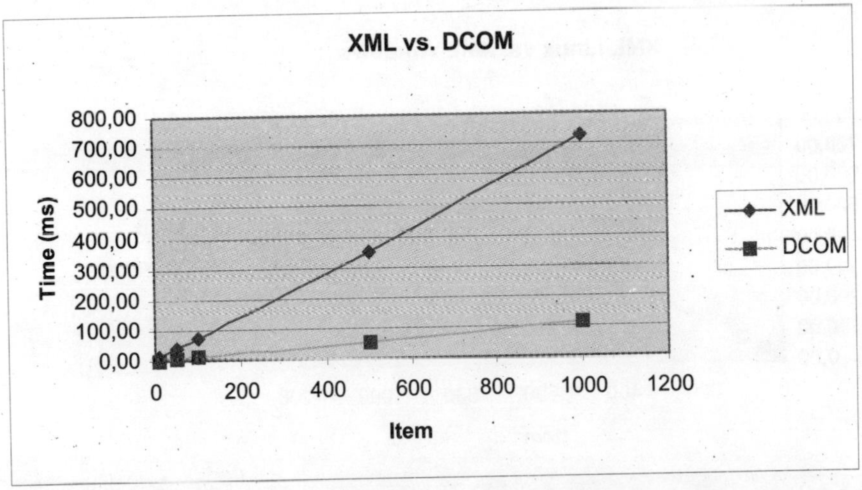

Figure 4.53: *Results for the measurement with OPC DCOM DA and OPC XML DA Server*

The first chart (**Figure 4.53**) shows the values measured for an OPC XML DA application and an OPC DCOM DA application. The individual values are listed in the table below.

	10	50	100	500	1000
OPC DCOM DA	2,15	7,03	12,39	54,87	110,33
OPC XML DA	12,10	37,31	73,71	352,66	730,01

The server and client each ran on different computers with a Windows operating system. The client was implemented in .NET. The server was programmed using the Softing C++ toolkit. The data exchange via OPC XML DA is slower by the factor of 6 than the data exchange via OPC DCOM DA. The values achieved, for example 730 ms for reading 1000 UI_4 values, are certainly acceptable for many applications.

OPC XML DA can be implemented on various platforms. Are there any difference in the performance, then? The chart (**Figure 4.54**) illustrates the comparison of measurements using an OPC XML DA server on Windows and Red Hat Linux. The computer that was used here was the same in both measurements.

There are no differences in the server performance. You will find the detailed values in the table below.

	10	50	100	500	1000
OPC XML DA Linux	11,07	37,95	71,12	325,68	704,12
OPC XML DA Windows	12,10	37,31	73,71	352,66	730,01

OPC XML DA applications can also run on the Internet. The last test was conducted to determine the corresponding effects on the performance. The chart shows the different values measured for access to the OPC XML DA server.

Access via the Internet was performed using a PC with Windows 2000, a 866 MHz processor and DSL connection. The server ran on a PowerPC 855T (Motorola) with

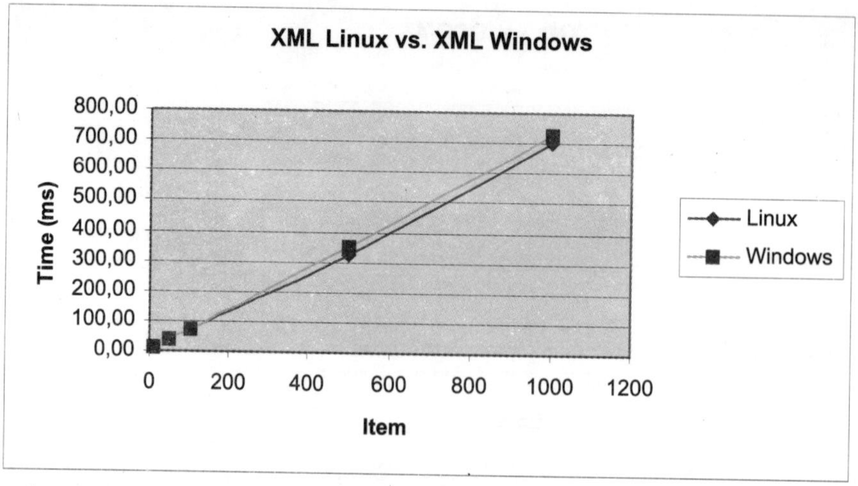

Figure 4.54: Results for the measurement with OPC XML DA Server on Windows and Linux platform

50 MHz. The computer was hooked up via a stub line behind the firewall and not connected directly to the intranet. Access to the computer was made from a client on the intranet to the Web service on the Internet. Since the firewall detected that the receiver and transmitter were in the same domain, the call was returned immediately. The absolute values for intranet access must not be compared with the above absolute values since a different processor architecture and, above all, a substantially slower processor were used. What is important is the absolute values for access via the Internet and the difference between access via the Internet and on the intranet.

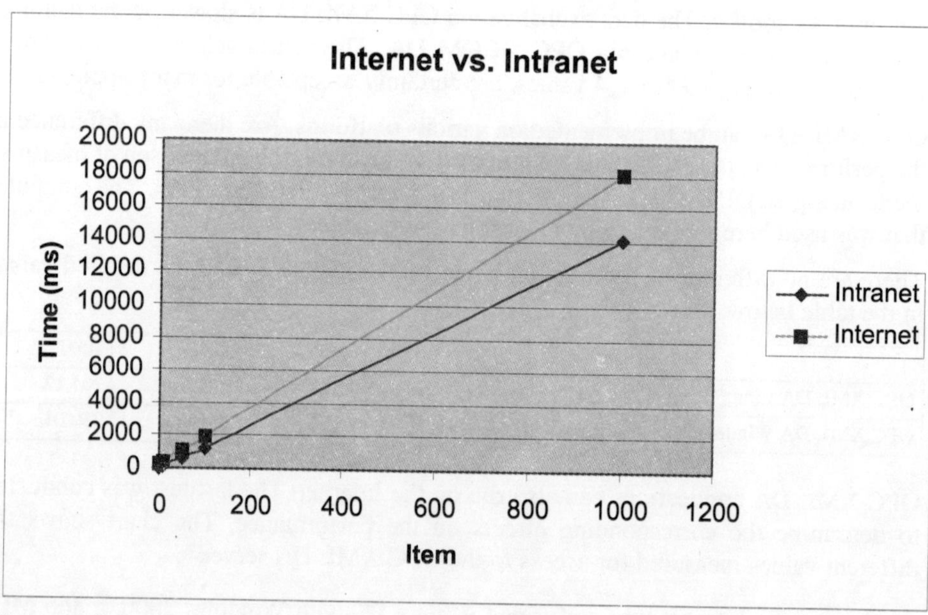

Figure 4.55: Access to an OPC XML DA Service via Internet and Intranet

You will find the values listed in detail in the table.

	1	10	50	100	1000
Intranet	40.2	149,2	623,5	1231,9	14134,4
Internet	271,95	417,88	1109,72	1923,29	18078

The relative difference decreases with an increasing number of items, i.e. with an increasing amount of data. Nevertheless, the values gathered for the Internet are somewhat surprising. 10 values are available in less than 1 s, 50 values in little more than 2 s and even 1000 values after 18 s.

The second table shows the amount of data exchanged between the client and service, and vice versa. What is also interesting is that the relative difference between intranet and Internet access decreases when the amount of data increases.

Item	1	10	50	100	1000
Data snd Client (byte)	739	1082	2643	4594	40596
Data rcv Client (byte)	795	1958	7138	13640	131571

The Web service is available at www.softing.com/webservices. Try access yourself. The **Figure 4.56** illustrates access using a sample client.

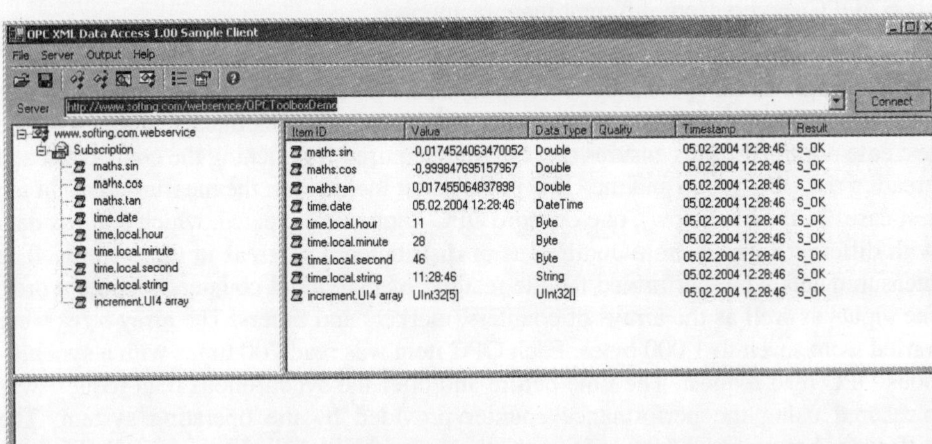

Figure 4.56: Access to an OPC XML DA Service via Internet

4.5.3 Performance Tests of OPC in Real-World Applications

The performance tests described in sections 4.5.1 and 4.5.2 provide valuable information on the absolute performance capabilities of OPC. They also show how the performance depends on the way the OPC communication is structured. In these tests, all influences from the acquisition of the data via a communication protocol were intentionally disregarded. In other words, the performance was tested for OPC components that do not perform real functions. In contrast, the performance test described in this section analyzes the actual performance of commonly used standard OPC servers from eight different manufacturers. OPC servers which allow access to Siemens S7 controllers via Ethernet were chosen for the test. The times for read access to one or more Siemens S7 controllers over the OPC interface were measured.

The measurements were performed on a 2.4 GHz computer with 256MB RAM and Windows 2000 operating system. Three S7-400 controllers with CP 443-1 communication processors were used in the tests. PLC 1 was connected to the test computer via 100-Mbit Ethernet, PLC 2 and 3 were connected via 10-Mbit Ethernet.

Table 4.3: *Specification of the three S7 test controllers*

PLC no.	CPU	CP	Ethernet in Mbits/s
PLC 1	6ES7 414 2XG03 – 0AB0	443-1EX11-0XE0	100
PLC 2	6ES7 414 1XG02 – 0AB0	443-1EX00-0XE0	10
PLC 3	6ES7 413 1XG00 – 0AB0	443-1EX00-0XE0	10

To measure the performance, the current product versions of the S7 OPC servers from the following manufacturers were tested: Applicom, IBH, INAT, Kepware, Langner, Siemens, Softing, and Wonderware. Two of the eight OPC servers did not meet all the requirements for the test (e.g. the array data type was not supported, or a value from the server cache was returned for the read request "Read from Device" without getting a new value from the device). The test results detailed later in this section thus cover six S7 OPC servers from different manufacturers.

An OPC client was developed as the test client on the basis of the Softing OPC Toolbox. This client opens a connection to the OPC server to be tested and creates an OPC group and the OPC items according to the test cases defined in Table 4.4 (see test case no. 6–10). This ensures that the time required for opening the connection and creating the OPC group and the OPC items is not included in the measurement. In ten test cases (see table below), one or more OPC items were created, which address data with different formats from contiguous or distributed data areas in the controls. The measurements were performed for the reading of individual counters (double word), the inputs as well as the arrays of counters, markers and timers. The array sizes were varied from seven to 1,000 bytes. Each OPC item was read 200 times with a synchronous OPC read request. The time before and after the synchronous read request was measured using the performance counter provided by the operating system. The difference between the time before and the time after the read request indicates how long the read operation actually took. A mean value was determined from the 200 measurements.

Table 4.4: *Definition of the 10 test cases*

Test case	Type of read access	(Size in bytes)	PLC
1	DB200.DBD0	Data block (DB): 200; one data word; (counter)	PLC 1
2	DB200.DBB100[1..1000]	DB: 200; one byte; offset: 100 array size: 1000	PLC 1
	DB200.DBD0	DB: 200; one data word (counter)	
3	DB200.DBB100[1..90]	DB: 200; one byte; offset: 100 array size: 90	PLC 1
	DB201.DBB100[1..90]	DB: 201; one byte; offset: 100 array size: 90	

Test case	Type of read access	(Size in bytes)	PLC
	DB202.DBB100[1..90]	DB: 202; one byte; offset: 100 array size: 90	
	DB203.DBB100[1..90]	DB: 203; one byte; offset: 100 array size: 90	
	DB200.DBD0	DB: 200; one data word (counter)	
4	DB200.DBB100[1..500]	DB: 200; one byte; offset: 100 array size: 500	PLC 1
	DB201.DBB100[1..500]	DB: 201; one byte; offset: 100 array size: 500	
	DB200.DBD0	DB: 200; one data word (counter)	
5	DB200.DBB100[1..50]	DB: 200; one byte; offset: 100 array size: 50	PLC 1
	EB0[1..8]	Input byte; array size: 7	
	MB0[1..30]	Marker byte; array size: 30	
	T0[1..10]	One timer; array size: 10	
	Z0[1..10]	One counter; array size: 10	
	DB200.DBD0	DB: 1; one data word (counter)	
6–10	The test cases 1–5 on two S7s		PLC 2 and PLC 3

As expected, the measurement results show that the data length, that is the number of bytes read, has an influence on the measured time. The transmission time is also affected by the degree to which the data to be read are "distributed" in the controller's memory. The OPC servers with the highest performance needed eight to nine milliseconds to read a data double word with a data length of four bytes from a PLC (test case 1). The same OPC servers transmitted 1,000 contiguous bytes (test case 2) from the controls to the OPC client in 35–40 milliseconds. From a controller with OPC items of the array type (test case 3), these OPC servers needed approx. 16–17 milli-

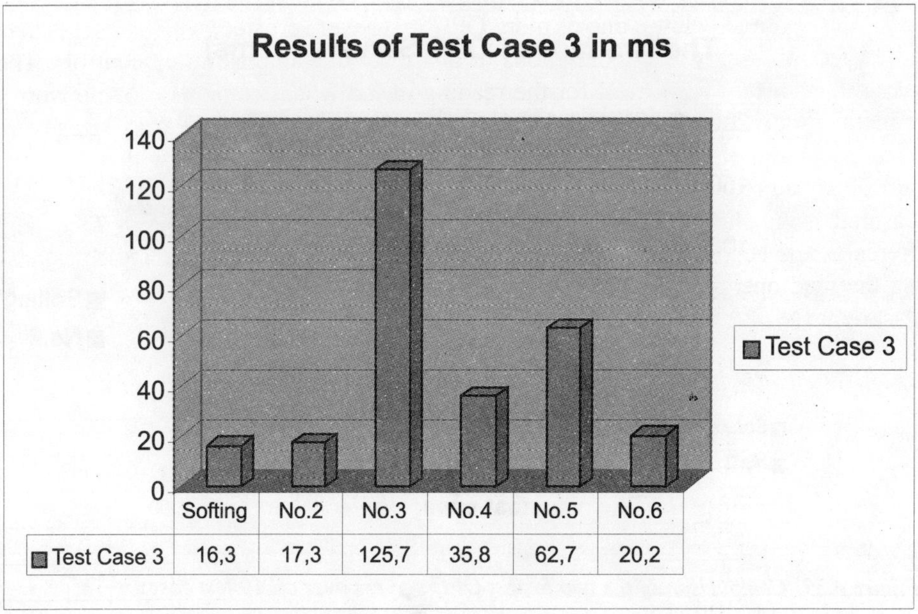

Figure 4.57: *Measurement results for test case no. 3*

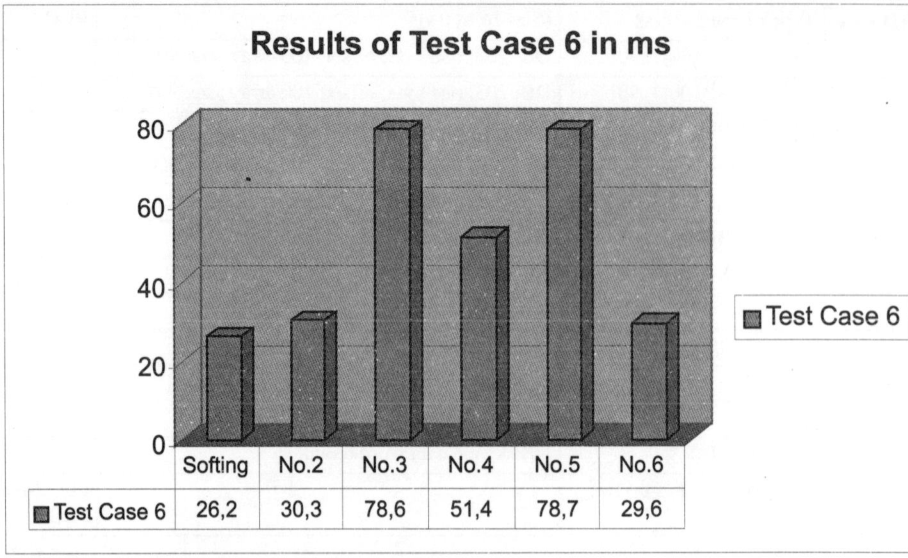

	Softing	No.2	No.3	No.4	No.5	No.6
Test Case 6	26,2	30,3	78,6	51,4	78,7	29,6

Figure 4.58: Measurement results for test case no. 6

seconds to read from four distributed data areas with a length of 90 bytes each. The more the data were distributed over the controller, the more the measurement results obtained for the different OPC servers varied. This is due to the fact that, depending on the number of OPC items, the configured update rate and the different address ranges in the control, a more or less complex algorithm is implemented to ensure optimum data transmission.

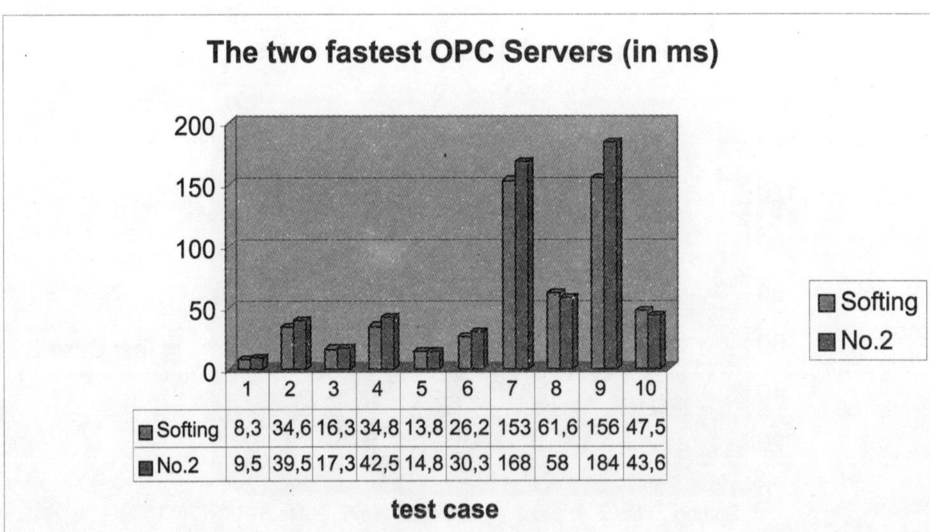

	1	2	3	4	5	6	7	8	9	10
Softing	8,3	34,6	16,3	34,8	13,8	26,2	153	61,6	156	47,5
No.2	9,5	39,5	17,3	42,5	14,8	30,3	168	58	184	43,6

Figure 4.59: Comparison of the two fastest OPC servers over all 10 test cases

The two figures above show examples of the test results obtained for all six OPC servers in the test cases three and six.

Figure 4.59 shows the results of the two fastest OPC servers for all 10 test cases. The S7 OPC server from Softing AG on average is approx. 10% faster than the second fastest product from a different manufacturer.

Softing AG offers the shortest access times in 8 of 10 test cases and is thus the fastest S7 OPC server in its class.

4.5.4 Distributed applications on the Internet with OPC

With the availability of DCOM it is obvious that OPC is used in distributed applications which not only exceed the boundaries of the PC but also those of administrative management units (domains).

In this case, users need to deal with products such as firewalls, the Internet, and Virtual Private Networks. This section will discuss this issue under the aspect of the use of OPC in distributed applications. **Figure 4.60** shows a possible structure and the components involved.

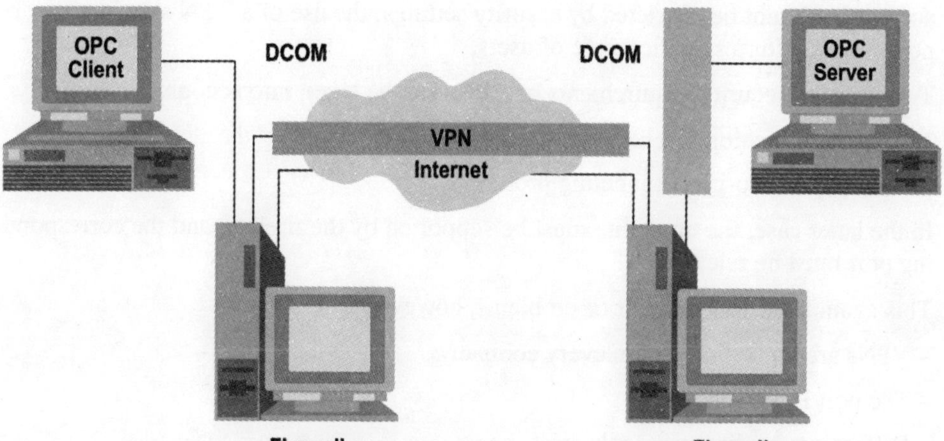

***Figure 4.60:** Distributed applications on the Internet with OPC*

Firewalls are installed at the interface between the Internet and the intranet. They are intended for monitoring and, if required, restricting external access to internal resources. Firewalls can also be used for limiting access to external resources.

There are three ways in which firewalls can perform their tasks:

1. Only specific ports and thus services are released for transit traffic.
2. Only the firewall can be reached from the outside at an IP address that translates all incoming calls into internal addresses (also known as Network Address Translation).
3. Proxies that check and extend incoming calls are installed for various tasks.

Internet applications have fixed ports over which they access PCs (HTTP, for example, always runs over port 80). A corresponding application is waiting at the corresponding port. Ports are assigned to DCOM applications dynamically and during runtime. Clients identify the port assigned to a DCOM server via a query to the Service Control Manager.

What settings need to be selected to be able to run DCOM via the firewall?

1. No address translation is permitted as DCOM saves IP addresses. If the client cannot connect to a PC under the specified address, no data will be exchanged.
2. The ports over which DCOM is to be run are negotiated dynamically during runtime. The range can be restricted, however, by means of settings.

Asking a system administrator will yield the response that these possibilities exist only theoretically. The Network Address Translation will not be switched off, and releasing the ports is also not feasible as this may otherwise cause gaps in the company's security shield. The next possible solution would be the use of Virtual Private Networks.

Virtual Private Networks consist of a private address range in public networks. Within VPNs, internal information is transmitted securely over public data channels. The access should not be restricted by security settings; the use of a VPN should rather be possible only for a specific range of users.

Two types of security requirements of VPNs can be taken into account:

1. Use of encryption mechanisms.
2. Use of point-to-point tunneling protocols.

In the latter case, the tunneling must be supported by the firewall and the corresponding port must be released.

This again leads to a number of problems, however:

- VPNs are not set up within every company.
- The port is not released.
- Different tunneling protocols exist.

What other possibility is there?

HTTP is used as the transport protocol between the client and server, and the OPC Servers are in a separate branch of the network behind the firewall. The firewall must not be connected to the intranet. Every user is permitted to launch the server. Other resources, however, are protected from being accessed according to the *OPC Security Specification*. The client now needs to use private credentials to be granted access.

This possibility can be used if:

- DCOM is mapped on HTTP over a gateway; however, this is a proprietary Microsoft solution.
- A data exchange is performed as defined by the OPC XML DA specification.
- DCOM calls are mapped on SOAP.

4.6 Summary

The OPC products and examples of applications presented in this chapter display the versatility in use of OPC today. OPC products are by no means used only in production plants for the operation and monitoring of processes. The connection of PC-based systems for visualization, acquisition of measured data, or control to the periphery is certainly still the “traditional” constellation in which OPC is applied. SCADA systems such as WinCC from Siemens, programs for the acquisition of measured data such as DIA*dem*® from National Instruments, or control systems such as 4CONTROL from Softing have been introduced as examples in this section. How widely OPC has been accepted in these areas of automation is also reflected by initiatives like those of the drive manufacturers in the Drivecom User Group or activities of the Buhler Group. TetraPak and Krupp Polysius are examples of end users that employ OPC in totally different industries. OPC servers like the LonWorks OPC server from who or the DTS OPC server from Softing are examples, how OPC is used as a standard in building automation resp. how applications in the automotive industry can be combined with applications in the automation industry using OPC. In their descriptions and in the reports of their experiences, the advantages of OPC are illustrated very comprehensively. How the use of standardized interfaces saves time and costs becomes very clear. Nearly any number of examples can be added to those of the companies mentioned – examples from food processing, mechanical engineering, chemistry, steel, in fact all industries in which automation technology is employed. The product catalog of the OPC Foundation provides an overview.

Apart from the “traditional” scenarios of application examples such as the one from Industrial and Financial Systems and from SAP show the advantages offered by an OPC interface in ERP systems. The IFS as well as the SAP company's ERP system illustrates very clearly how OPC connects the automation level with the business level. OPC today plays a key role in vertical integration concepts.

With increasing availability of OPC XML-DA products the use of the Internet and the Web technologies involved will grow even faster. The possibilities of connecting OPC products over the Internet and using OPC also on non Windows platforms will lead to further areas of application, especially in the area of remote maintenance and remote diagnosis. The work in the new *OPC Unified Architecture* Working Group in this regard certainly constitutes a flanking measure which makes it imaginable that DCOM-based OPC products will soon be communicating with applications on completely different operating systems that are not “DCOM-compatible.”

That OPC does not need to fear comparison in terms of performance is shown by the figures from the performance tests of section 4.5.1. The results of the performance assessments of OPC XML-DA connections show, that beside the huge advantages of Internet capability and platform independence flow rate is the “bottle neck”. The fast multiplication of CPU frequencies on the one hand and using standardized binary XML transmission on the other hand indicate that this disadvantage will be eliminated in a foreseeable time.

5 Summary and Outlook

Since the publication of the first and second editions of this book, the success story of OPC has continued at an amazing pace. The amount of over 7,000 copies sold of both the German and English editions in the four years since first publication has shown the tremendous interest in OPC.

The rapid acceptance and widespread application of OPC technology in a vast number of areas in automation technology is unique. Many standards, e.g. regarding fieldbuses in industrial communication or in PLC programming, took a decade or even longer to become established. OPC succeeded in only four years in introducing an industrial standard without which automation technology today, nine years later, would be unimaginable. The specifications released until now cover a wide variety of application areas, from the fast data transfer (*Data Access*) and the processing of large amounts of historical data (*Historical Data Access*) to the acquisition and acknowledgment of volatile and critical events (*Alarms and Events*).

The availability of Data eXchange and OPC XML-DA opens up further areas of application in vertical integration and peer-to-peer interaction. On the basis of the specifications, clients and servers can run in one process on one PC (InProc Server) or in two processes (OutProc Server) on PCs that can be close together or far apart. One or several clients can access a server. A client, in turn, can communicate with one or more servers.

Most of the OPC products installed to date (October 2005) are still based on Microsoft DCOM technology. Even though Microsoft platforms will continue to support DCOM in the next few years, Microsoft .NET as the successor to DCOM for the development of distributed object-oriented applications will be more and more common in the next few years. Numerous .NET developments are currently in progress.

The *OPC XML-DA Specification* takes this new situation into account by defining the programming interface as a Web service. With Web services based communication, OPC has achieved platform independence. OPC communication will thus continue to be possible over the Internet, across firewall boundaries.

OPC technology today is a standard that enables global interoperability and provides the required information to any authorized person anywhere and anytime.

The start of the new *OPC Unified Architecture* initiative, the definition of an interoperability platform as well as numerous collaborations with other standardization organizations promise exciting new possibilities with OPC.

There are a variety of products which translate the OPC specifications into reality. The scope reaches from sample clients and servers to tools for creating OPC components and servers for a wide variety of communication systems and devices. A multitude of visualization systems, systems for data acquisition and diagnosis have been equipped with OPC Client interfaces. These products are used in a vast number of actual applications. A large variety of programming languages and development environments are available for the creation of OPC components. OPC products can be used on a wide range of hardware and software platforms.

Solutions for the compliance test of *Data Access*, *Alarms and Events*, *Historical Data Access,* and *OPC XML-DA Servers* are available. The use of successfully-tested products raises the confidence of end users in product quality and interoperability.

A non-profit organization, the OPC Foundation is in charge of marketing and developing the technology.

These points characterize the current actual state described in the book. One question still remains to be answered – What does OPC's future look like?

Future development of OPC

Most OPC specifications are based on Microsoft DCOM technology. As explained in chapters 2.4 (XML-DA), 2.5 (DX) and 3.4, the *OPC XML-DA* and *OPC DX* specifications have taken OPC technology a decisive step further – to the integration of XML and Web services. The OPC Foundation has thereby implemented the technological basis defined by Microsoft with .NET as the successor to DCOM for the development of distributed object-oriented applications. Although the first Web services based OPC XML-DA products are already being employed in industrial applications and although a trend is emerging toward the use of the OPC Foundation's .NET API, this only marks the beginning of use of the "new OPC generation". The upcoming release of the *OPC Unified Architecture Specification* is expected to generate a substantial increase in XML and Web services based OPC products.

The vision behind OPC Unified Architecture

The OPC Unified Architecture is the new OPC architecture and series of specifications that is the vision behind secure reliable interoperability for moving data/information from the factory floor to the enterprise. It takes the existing OPC specifications and ties them all together leveraging web services as the key technology to deploy the new architecture. The OPC Unified Architecture provides the framework for reliability which is a very important issue of moving information between applications in the enterprise applications space. The side effect of the work that is being done with respect to the enterprise is that the control systems will benefit from the reliable delivery mechanisms that are laced with security that is the framework of the OPC Unified Architecture.

Furthermore the OPC Unified Architecture also leverages the collaboration with the other industry standards organizations inclusive of ISA SP95, ISA SP99, IEC, OAGi and MIMOSA, where the OPC Unified Architecture is the how for moving the what, where the what is defined to be the information models from the other industry standards organizations listed above. The OPC Unified Architecture provides a base

foundation of services for the discovery, read/write and subscription capability for moving information that is described in the companion specifications of the other industry standards organizations that focus on information modeling. The services that the OPC Unified Architecture provides for secure reliable interoperability also are the result of collaboration with other Web services standards organizations that span other industries including retail, financial, medical and security.

In addition the OPC Unified Architecture provides the wrapper technology that supports legacy DCOM based products to be easily enabled to the new OPC Unified Architecture. The plan is such that the OPC Unified Architecture wrapper could be installed for both the OPC client and OPC server products, such that the reliability problems that have existed between typical OPC client and server products currently deployed in the field would magically go away with the installation the OPC Unified Architecture wrapper components. This allow existing OPC DCOM based products that are already installed on the factory floor to not require any modifications to become OPC Unified Architecture enabled, which is a significant benefit a sense the possibility that existing OPC DCOM based products that may have been installed in the late 1990s can not easily be modified to support the new OPC Unified Architecture services, because the company might be out of business that originally developed the technology.

There is also a close collaboration with Microsoft providing the key technology to enable rapid employment of the OPC Unified Architecture in both control and IT space. Although the OPC Unified Architecture is platform independent, products that be deployed on Windows based operating systems will serve well to be developed on the Microsoft technology, that provides the foundation for rapid development of OPC Unified Architecture products. The Microsoft Windows Vista will be the first unveiling of many of the core services that are the foundation of the OPC Unified Architecture.

In summary, the OPC Unified Architecture is the end-user driven architecture and series of specifications and components to address secure reliable interoperability. The OPC foundation serves as the conduit for the end-user community voice to the vendor to develop products that meet and exceed the end-user requirements for secure reliable interoperability.

The speed at which *OPC UA* will find acceptance will largely depend on the solution to the performance requirements. Breakthrough of the new OPC generation is anticipated on reaching the goal set by the *OPC UA* working group to define a binary XML protocol for achieving a data transfer rate that is equivalent to DCOM, if not faster. The envisaged performance would offer various advantages to the manufacturers and users of OPC products: from providing and using open OPC communication also on non-Microsoft operating systems to embedding OPC products in field devices, controllers, etc., and communicating via the Internet independently of the manufacturer. No distinction would then need to be made as to whether the data read are process values from a Data Access server, alarms from an Alarms and Events server or mean values from a Historical Data Access server. The unification of the OPC namespace in *OPC UA* will further simplify the use of even a large number of different OPC components. Signs already indicate that OPC technology will become even more widespread. New collaborations between the OPC Foundation and other standardization organizations as well as the adoption of OPC in recommendations, directives and specifications in almost all areas of the industry point to the fact that OPC is rapidly entering new application areas and branches of industry that have not been using OPC so far. The OPC Foundation aims at making OPC 'the' interoperability platform that defines how data and information is exchanged worldwide, independently of the specific manufacturers, industries and platforms.

A few personal words from the authors

We hope that reading this book will help you to better understand OPC technology and many related areas. Maybe you will influence further development of OPC yourself in the future. We would be glad if you did.

Bibliography

Literature used in making this book

OPC Overview. Version 1.0, October 27, 1998, OPC Foundation

OPC Common Definitions and Interfaces Version 1.0, October 27, 1998, OPC Foundation

OPC Commands Version 1.00, April 2004, OPC Foundation

OPC Complex Data Version 1.00, December 2003, OPC Foundation

OPC DX Version 1.00, March 2003, OPC Foundation

OPC Specification Version 1.0, September 1996, OPC Foundation

OPC Data Access Specification Version 1.0A, September 1997, OPC Foundation

OPC Data Access Custom Interface Standard Version 2.05, June 2002, OPC Foundation

OPC Data Access Custom Interface Standard Version 3.00, March 2003, OPC Foundation

OPC Data Access Automation Interface Standard Version 2.02, February 1999, OPC Foundation

OPC Alarms and Events Custom Interface Standard Version 1.10, November 1999, OPC Foundation

OPC Alarms and Events Automation Interface Standard Version 1.01, December 15, 1999, OPC Foundation

OPC Historical Data Access Custom Interface Standard Version 1.20, December 2003, OPC Foundation

OPC Historical Data Access Automation Interface Standard Version 1.0, January 2001, OPC Foundation

OPC Batch Custom Interface Specification, Version 2.0, July 2001, OPC Foundation

OPC Batch Automation Interface Specification, Version 1.0, July 2001, OPC Foundation

OPC Security Custom Interface Version 1.0, July 2001, OPC Foundation

OPC XML-DA Specification Version 1.01, December 2004, OPC Foundation

Technical Agreement Summary Version 1.0 Draft, February 2002, OPC Foundation

Zertifizierung von OPC-Komponenten. SPS Magazin Ausgabe 3+4/2000 S. 34-35 (German only)

Inside Distributed COM. Guy Eddon, Henry Eddon. Microsoft Press. ISBN 1-57231-849-X. 1998

COM Internet Services. Marc Levy. Microsoft Corporation. April 23, 1999. MSDN-Library

Using Distributed COM with Firewalls. Michael Nelson. June 20, 1998. MSDN-Library

Untersuchungen von Performance und Security beim Einsatz von OPC. Diplomarbeit von Alexander Stamm an der Fachhochschule Frankfurt FB 04 Diplomarbeit E 00133. 2000 (German only)

List of figures

List of tables

Index